Heterogeneous Photocatalysis

Wiley Series in Photoscience and Photoengineering

Executive Editorial Board

Volume 1
Surface Photochemistry

Edited by

Masakazu Anpo
University of Osaka Prefecture,
Osaka, Japan

Volume 2
Homogeneous Photocatalysis

Edited by

Michel Chanon
Université de Droit,
D'Economie et des Sciences d'Aix-Marseille,
Marseille, France

Volume 3
Heterogeneous Photocatalysis

Edited by

Mario Schiavello
Università degli Studi di Palermo,
Palermo, Italy

Heterogeneous Photocatalysis

Edited by
M. Schiavello
Università degli Studi di Palermo, Palermo, Italy

WILEY SERIES IN PHOTOSCIENCE AND PHOTOENGINEERING VOLUME 3

JOHN WILEY & SONS
Chichester • New York • Weinheim • Brisbane • Singapore • Toronto

Chemistry Library

Other Wiley Editorial Offices

John Wiley & Sons, Inc., 605 Third Avenue,
New York, NY 10158-0012, USA

WILEY-VCH Verlag GmbH, Pappelallee 3,
D-69469 Weinheim, Germany

Jacaranda Wiley Ltd, 33 Park Road, Milton,
Queensland 4064, Australia

John Wiley & Sons (Asia) Pte Ltd, 2 Clementi Loop #02-01,
Jin Xing Distripark, Singapore 129809

John Wiley & Sons (Canada) Ltd, 22 Worcester Road,
Rexdale, Ontario M9W 1L1, Canada

Library of Congress Cataloging-in-Publication Data
Heterogeneous photocatalysis / edited by M. Schiavello.
     p. cm. — (Wiley series in photoscience and photoengineering
; v. 3)
  Includes bibliographical references and index.
  ISBN 0 471 96754 8 (alk. paper)
  1. Photocatalysis. 2. Heterogeneous catalysis. I. Schiavello,
Mario, 1935– . II. Series.
QD716.P45H48 1997
541.3'95—DC21
                              97-13275
                              CIP

British Library Cataloguing in Publication Data

A catalogue record for this book is available from the British Library

ISBN 0 471 96754 8

Typeset in 10/12pt Times by Techset Composition Ltd, Salisbury
Printed and bound in Great Britain by Biddles, Guildford and King's Lynn
This book is printed on acid-free paper responsibly manufactured from sustainable forestation,
for which at least two trees are planted for each one used for paper production.

Contents

Series Preface

Looking into the history of chemistry, one of the fascinating facts is that discoveries and developments which helped to shape today's civilization and technical standards have frequently been made on a totally empirical basis and it has usually been quite some time afterwards that fundamental research developed the means for understanding the implications of chemical structures, interactions, reactivities and physical characteristics.

Despite the fact that light plays a key role in our daily life, photochemistry found only limited recognition as a proper domain of chemical sciences and engineering. For most chemists, photochemistry has been a playground of academic study. But considering technical developments in microelectronics, informatics, (dental) medicine, materials, fine and bulk chemicals, etc., for example, our world would be different if the industries concerned had not found the relevant results by using light as a reagent in rather complex reaction systems; and academic research has been in most cases bypassed by industrial success.

With the series 'Photoscience and Photoengineering', we aim to foster interaction between fundamental research and technical development. Topics of the volumes will be carefully selected depending on the impact of new developments and the knowledge of the related basic principles. Therefore, these books will not be ensembles of loosely related reviews, but self-contained accounts on specific areas.

I sincerely hope that our approach will lead to a better understanding and a closer interaction between scientists and engineers of many disciplines.

Karlsruhe, September 9, 1995
Prof. Dr André M. Braun
Chairman of the Board of Editors

Preface

Photocatalysis is a fast growing area with respect to both applied and theoretical research. This volume outlines the fundamental principles of heterogeneous photocatalysis as well as presenting a number of case studies which illustrate how these principles can be applied to engineering problems.

A volume on Homogeneous Photocatalysis has already been published in this series.

Chapter 1 of this volume deals with solid state chemistry and Chapter 2 deals with the fundamentals of surface chemistry. In Chapter 3, the basic principles of photo-adsorption and photo-desorption processes are described. Chapter 4 outlines the fundamentals of thermodynamics and kinetics to enable the reader to understand why and how heterogeneous photocatalytic processes occur. Chapter 5 describes reduction photoprocesses. Chapter 6 concludes the volume with a discussion of the main engineering problems involved in the application of photocatalytic processes.

Contributors

M. Anpo
Dept. of Applied Chemistry, University of Osaka Prefecture, Sakai City, Osaka 593, Japan

A. Atrei
Università di Firenze, Dip di Chimica, Laboratorio di Chimica Fisica delle Interfasi, via Cavour 82 50129 Firenze, Italy

V. Augugliaro
Università degli Studi di Palermo, Viale delle Scienze, 90128 Palermo, Italy

J. Augustynski
Université de Genève, Dept. de Chimie Minérale, Analytiques, Applique, 30 Quai E Ansermet 1211, Genève, Switzerland

R. I. Bickley
University of Bradford, Dept. of Chemical Engineering, Chemistry and Chem. Technology, Bradford BD7 1DP

V. Loddo
Università degli Studi di Palermo, Dipartimento di Ingegneria Chimica dei Processi e dei Materiali, Viale Delle Scienze, 90128 Palermo, Italy

L. Palmisano
Università degli Studi di Palermo, Viale delle Scienze, 90128 Palermo, Italy

G. Rovida
Università di Firenze, Dip. di Chimica, Laboratorio di Chimica Fisica delle Interfasi, via Cavour 82, 50129 Firenze, Italy

M. Schiavello
Università degli Studi di Palermo, Dip. di Ingegneria Chimica dei Processi e dei Materiali, Viale Delle Scienze, 90128 Palermo, Italy

A. Sclafani
Università degli Studi di Palermo, Viale delle Scienze, 90128 Palermo, Italy

M. Voïnov
Université de Genève, Dept. de Chimie Minérale, Analytiques, Applique, 30 Quai E Ansermet, 1211 Genève, Switzerland

H. Yamashita
Dept. of Applied Chemistry, University of Osaka Prefecture, Sakai City, Osaka 593, Japan

1 Introduction to the Physics of Semiconductor Photocatalysts

M. VOÏNOV
Ecole d'Ingénieurs de Genève, Geneva, Switzerland
and
J. AUGUSTYNSKI
Université de Genève, Geneva, Switzerland

Heterogeneous Photocatalysis, Edited by M. Schiavello
© 1997 John Wiley & Sons Ltd.

1 INTRODUCTION

The exact definition of the term *heterogeneous photocatalysis* is a difficult one, especially as in many cases the detailed mechanism of the ongoing reactions is uncertain. However, in all cases, such a reaction scheme implies the previous formation of an interface between, in general, a solid photocatalyst and a liquid or a gas phase containing the reactants and/or the products of the photoreaction. The common case is that of a light-absorbing semiconductor in contact with either a liquid or a gas phase. However, there is in principle no reason to exclude from the field of heterogeneous photocatalysis the reactions undergone by the species, present at the surface of a metal such as silver, following the photoexcitation of surface plasmons in the metal and the subsequent photoelectron emission. In a more general way, the entire, relatively new branch—the photochemistry occurring at adsorbate/metal interfaces—also belongs to the field of heterogeneous photo-catalysis. The latter example illustrates the important role played by the adsorption of reactants, leading to a change in their chemical and optical properties. The series of events following the illumination of a metal–liquid or a metal–gas interface may be initiated either by light absorption by the metal, leading subsequently to the activation of the adsorbate, or by direct excitation of the adsorbate which is then quenched by the metal.

Both these mechanisms are in fact also encountered in the case of semi-conductor–liquid and semiconductor–gas interfaces and may even sometimes operate simultaneously. This can, in particular, be expected when the reactants adsorbed at the semiconductor surface are illuminated with a large spectrum UV light. Thus, the formal distinction between so-called sensitized photoreactions, where the primary step consists in the photoexcitation of a catalyst and is followed by the charge transfer into a ground state reactant, on the one hand, and catalysed photoreactions, where the initial photoexcitation occurs in the adsorbed reactant which then transfers electron or energy to the catalyst, on the other hand, is fre-quently quite aleatory.

The activity and selectivity of a solid photocatalyst results from a particular combination of bulk and surface properties. The present chapter is devoted to a general description of basic principles which govern the bulk properties of the most widely employed class of photocatalysts, which consists of semiconductors. However, some special features associated with very small dimensions of semi-conductor clusters—the so-called quantum size effects—are not addressed here.

For irradiated massive semiconductor photocatalysts, the number of photons absorbed in the first two or three atomic surface layers is usually much smaller than the number absorbed in the bulk. Then, excited states created in the bulk of the photocatalyst must somehow reach the surface to participate in desired chemical reactions. In this situation, it is tempting to use solid state concepts such as elec-tronic allowed and prohibited energy bands, Fermi levels, band bending, electrons and holes mobilities, density of states, etc. to describe their behaviour.

However, one should always keep in mind that all these concepts and theories are strictly applicable to covalent semiconductor crystals. Stretching them to describe properties of typical ionic compounds generally used in photocatalysis should be done with care. For instance, many physicists would be tempted to agree with Ziman [1] to the effect that 'the concept of valence band is probably meaningless in an ionic compound' as photoinjected holes interact so strongly with optical phonons that they are self-trapped and therefore hardly delocalised at all.

In the case of a finely dispersed catalyst, the number of surface atoms is no longer negligible compared with the number of bulk atoms and the proportion of photons absorbed in the surface layers of such catalysts becomes important. However, the extension of solid-state theories to surface atoms is slightly perverse because these theories very often use a Born–von Karman periodic boundary condition to present solutions [1–4]. This condition suppresses, for better or worse, all problems associated with surfaces or even heterojunctions. For small particles, the chemical bond or amorphous semiconductors theories are actually more relevant to surface atoms than the crystalline solid-state theory. Inside these small particles, electrons are very superficially described as wave-particles in a perfect, generally square or harmonic, potential well [5].

In the present chapter, some of the solid-state theory concepts and approximations are presented in so far as they appear relevant to photocatalysis.

2 THE ADIABATIC APPROXIMATION AND THE FRANCK–CONDON PRINCIPLE

Solid materials contain agitated nuclei and electrons interacting via electromagnetic photons. Electrons are much lighter than nuclei and consequently, at any given temperature, they move much faster than nuclei. Furthermore, the speed of the electromagnetic interaction between all these particles is the highest possible indeed. That is why, in the so-called adiabatic or Born–Oppenheimer approximation [6], it is considered that the energy and momentum of each electron adjust instantaneously to the nuclei movements. One obvious drawback of this approximation is that it suppresses important physical phenomena such as inelastic electron scattering by lattices, i.e. the ohmic behaviour of conductors. The adiabatic approximation does not imply that electrons are always in equilibrium with the lattice movements. Indeed, they may well adopt excited states that do not correspond to equilibrium nuclei positions. For instance, in solids, nuclei are vibrating around a mean position at frequencies of the order of 10^{13}–10^{14} Hz, and so the adsorption and emission of electromagnetic radiation having frequencies higher than 10^{14} Hz occurs without appreciable nuclei movement. This is known as the Franck–Condon principle and the higher the frequency of the absorbed photon the more valid it is. As we shall see, the Franck–Condon principle is not absolute and, in particular, it does not prohibit lattice vibrations (phonons) to participate in electronic radiative transitions inside solids. Such transitions, assisted by lattice motions, do occur in so-called indirect band gap materials with emission or absorption of phonons.

3   SPONTANEOUS AND STIMULATED TRANSITION

Granted that in atoms, molecules and materials, electrons can only have discrete energy values or levels, transitions between any two levels may be either spontaneous or stimulated by photons. But it should be borne in mind that, when a photon has exactly the right (resonance) energy to stimulate an electron to cause a transition to an upper energy level (photon absorption), it also has the right energy to stimulate (with the same probability) a downward transition (photon emission). The number of electrons in upper energy levels is then determined not only by the probability of spontaneous emission but also by the probability of stimulated emissions.

As shown by Einstein, the ratio of spontaneous to stimulated transition probability increases with the third power of the energy difference between levels.

Overall, to have an appreciable population of electrons in a photon-excited state, there must exist a mechanism which decreases the probability of spontaneous transition as the probability of stimulated absorption is equal to the probability of stimulated emission. Inside solids, two such mechanisms are, for example, decay (leakage or thermalisation) of excited electrons into traps and also phonon-assisted transitions.

4   ENERGY BANDS AND GAPS

There are basically two ways to introduce electron energy bands and gaps existing in solids. In both ways, electrons are considered to be travelling or stationary (complex) waves obeying linear differential equations so that their amplitudes can be added to obtain interference phenomena:

1. In the first approach, the one-electron wave functions are travelling complex plane waves in a periodic $V(r)$ potential. The periodicity is the crystalline periodicity.
2. In the second approach, the wave functions are obtained from individual stationary atomic orbitals extending over the neighbouring atoms and mixing in such a way that the resulting global wave function is antisymmetrical upon exchange of any two electrons between two allowed states.

4.1   TRAVELLING WAVES IN A PERIODIC POTENTIAL

The wave function of a free electron of wave vector $\mathbf{k}$ is

$$\Psi_k = \exp(i\mathbf{k} \cdot \mathbf{r}) \tag{1.1}$$

where $\mathbf{r}$ is a spatial vector.

Bloch's theorem [7] states that, in a periodic potential, the wave function for a free electron is obtained by adding wave functions differing only by reciprocal lattice vectors $\mathbf{G}$:

$$\Psi_k(\mathbf{r}) = \sum_k C_{\mathbf{k}+\mathbf{G}} \exp[i(\mathbf{k}+\mathbf{G}) \cdot \mathbf{r}] \tag{1.2}$$

where $C_{\mathbf{k}+\mathbf{G}}$ are suitable coefficients.

This is equivalent to:

$$\Psi_k = \exp(i\mathbf{k} \cdot \mathbf{r})\mathbf{U}(\mathbf{r}) \tag{1.3}$$

where $\mathbf{U}(\mathbf{r}) = \mathbf{U}(\mathbf{r}+\mathbf{b})$ and $\mathbf{b}$ is the spatial period of the potential.

The above Bloch wave functions can be also expressed as sums of travelling plane waves interfering to yield stationary states. Finally Bloch wave functions can be assembled into localised wave packets to represent electrons that propagate freely in materials.

The periodic potential $U(\mathbf{r})$ seen by the electron can also be developed as a sum:

$$U(\mathbf{r}) = \sum_G \mathbf{U}_G \exp(i\mathbf{G} \cdot \mathbf{r}) \tag{1.4}$$

where $\mathbf{G}$ is a reciprocal lattice vector.

It is also generally assumed that the sum converges rapidly which, in reality, is very unlikely. The only apparent justification for such an assumption is that it works rather well.

Putting these two sums in the appropriate time-independent Schrödinger equation, one obtains the three following desired features:

1. *singularities in the kinetic energy versus momentum relation* when the tip of the wave vector is on the surface of Brillouin zones, that is:

$$\mathbf{k} \cdot \tfrac{1}{2}\mathbf{G} = \tfrac{1}{4}\mathbf{G}^2 \tag{1.5}$$

   (Bragg diffraction condition);
2. *allowed energies bands and forbidden energies gaps*;
3. *mobilities and masses of carriers depending altogether on wave vector, band width, band gap and spatial direction*.

These results come directly from the equations written, so that the underlying physics is a bit hidden.

The *singularities in the kinetic energy/momentum relation* are due, in fact, to the reflection of plane travelling waves on the periodic potential; a plane wave travelling in one direction with a wave vector **k** is reflected with a wave vector **k** − **G**, so that when **k** approaches one half of a reciprocal lattice vector the superposition of the two waves of opposite wave vector $\frac{1}{2}$**G** and $-\frac{1}{2}$**G** and of the same amplitude produces two standing waves of the same spatial period.

The *existence of an energy gap*, when the wave vector is $\frac{1}{2}$**G**, is due to the fact that the two standing waves are spatially out of phase, producing electron concentration on lattice sites for one wave and electron concentration half way between lattice sites for the other (Figure 1.1).

A chemist may be reminded here of bonding and anti-bonding molecular orbitals. As the periodic electrical potential seen by the electron varies with crystallographic directions, so do the band gap, band width and energy E versus wave number k relations. The apparent *variation of the electric current carrier mass* is a consequence of its wave-like character. In classical dynamics the relation between kinetic energy, E, and momentum, p, is

$$E = \frac{p^2}{2m} \tag{1.6}$$

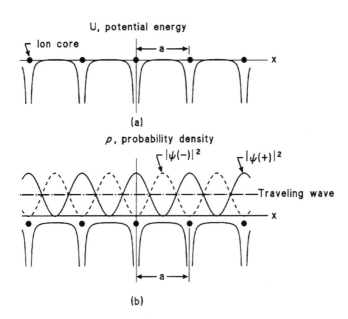

Figure 1.1.   (a) Variation of the potential energy of a conduction electron in the field of the cation cores of a linear periodic lattice. The standing wave $\Psi(+)$ piles up electronic charges on the cores of the positive ions thereby lowering the potential energy seen by a travelling wave. The standing wave $\Psi(-)$ piles up charges in the region between ions, thereby raising the potential energy. The difference between these raised and lowered allowed energies constitutes the band gap. (b) Distribution of probability density p for the two spatially out of phase standing waves at an energy versus wave vector discontinuity and also for a travelling wave (from reference [2], reproduced by permission of John Wiley & Sons, Inc.).

Because of the travelling wave aspect of moving particles this relation is also written as

$$E = \frac{\hbar^2 k^2}{2m} \tag{1.7}$$

where $\hbar$ is (Planck constant)$/2\pi$.

In both cases, the coefficient, m, is by definition the inertial mass of the wave-particle. For a non-relativistic classical free particle, the parabolic relation between kinetic energy and momentum delivers a constant mass. But in a periodic potential, there exist singularities in the relation between E and k, so that the parabolic relation is no longer obeyed, especially near the singularities. The coefficient m varies with the wave vector: because of this variation it is then called the effective mass, m^*, and defined as

$$m^* = \frac{\hbar^2}{\partial^2 E/\partial k^2} \tag{1.8}$$

The physical reason behind the apparent variation of m^*, at least near zone boundaries, is that for wave vector k approaching $\frac{1}{2}G$, any outside effort to increase the momentum of a particle (by application of an electrical field for instance) increases the amplitude of the reflected wave and therefore momentum transfer to the lattice at the cost of actual momentum transfer to the particle. It looks as though the increase of particle momentum due to outside influence is smaller than expected or that the mass is smaller than the rest mass. The effective mass, m^*, can even be formally negative near the top of the allowed bands.

Because $E(k)$ is orientation dependent, so is m^* which, in general, is written as a tensor, the components of which are

$$\left(\frac{1}{m^*}\right)_{ij} = \frac{1}{\hbar^2} \frac{\partial^2 E(k)}{\partial k_i \partial k_j} \tag{1.9}$$

An important result of the travelling-wave-in-a-periodic-potential approach is that the ratio of the effective mass to the rest mass increases with the magnitude of the band gap E_g and is inversely proportional to the band width W

$$m^* \div \frac{E_g}{W} \tag{1.10}$$

Therefore, narrow band gap materials will tend to have carriers with small effective masses and high mobilities. On the other hand, in narrow allowed bands, the effective masses are likely to be high with associated low mobilities. Highest effective masses are expected with narrow band-width and large band-gap materials, typically transition metals ionic compounds. These general trends are confirmed by the data listed in Table 1.1.

Another important result is that because the periodic potential varies obviously with the crystallographic directions, so do the band gap, band width and effective masses.

Table 1.1. Band gaps, mobilities and effective masses of charge carriers for selected semiconductor materials

Materials	Band gap (eV) at 300 K	Electron mobility (cm²/(V s)) 300 K	Hole mobility (cm²/(V s)) 300 K	Smallest m^*/m_0
Diamond	5.47	1 800	1200	h: 0.7
Si	1.12	1 350	480	e: 0.19
				h: 0.16
Ge	0.66	3 600	1800	e: 0.08
				h: 0.045
α-Sn	0.07	1 400	1200	
α-SiC	2.99	100	10–20	
AlSb	1.58	900	400	
GaSb	0.72	5 000	1000	e: 0.047
				h: 0.06
GaAs	1.42	8 000	300	e: 0.066
				h: 0.082
GaP	2.26	110	75	
InSb	0.17	80 000	450	e: 0.015
				h: 0.021
InAs	0.36	30 000	460	
CdS	2.42	340	50	
CdSe	1.7		800	
CdTe	1.56	1 050	100	
ZnO	3.35	200	180	
ZnS	3.68	165	5	
PbS	0.41	550	600	
PbSe		1020	930	
PbTe	0.31	2500	1000	
TiO₂₋ₓ (rutile)	3.03	0.1		e: 20–40
(anatase)	3.2	4–20		e: 1
BaTiO₃₋ₓ	2.5–3			e: 10
Cu₂O	2.1			e: 0.99
				h: 0.58
Mn₁₋ₓO		10	10–5	
LiₓMn₁₋ₓO			10–2–10–5	
AgCl		50		
SnO₂	3.6	240		e: 0.39

In spite of its success, a very important limitation of the above approach is that all essential features and, principally, the existence of band gaps, are due to the periodic electrical potential associated with crystallinity. But one knows of the existence of band gaps in amorphous semiconducting (Si) and insulating (Se photoconductor) materials [8,9]. Obviously a periodic potential can introduce a band gap, as we just saw, but it is not an absolute necessity. Other approaches may give essentially the same result.

4.2   COMBINATION OF NEAR NEIGHBOURING ATOMIC ORBITALS

In isolated atoms allowed electron energies are quantified in s,p,d,... atomic orbitals. But when atoms gather to form a solid, electrons of each atom tend to spread into neighbouring allowed orbitals. This is prohibited by the necessity of having antisymmetrical wave function (Pauli exclusion principle) for each pair of electrons. Consequently in a solid, the atomic orbitals of each type (for example s type) form a band of closely spaced (on an energy scale) allowed orbitals [10–12]. Inside a band the atomic orbitals keep their identities but the difference in energy between two such adjacent orbitals is generally much smaller than the thermal energy of phonons so that, in a first approximation, they form a continuum of allowed energies. As shown schematically in Figure 1.2, in each band, the difference in energy between the highest and the lowest allowed orbital depends on the distance between neighbouring atoms.

In addition to energy bands and gaps, the two following qualitative features, not obvious in the previous travelling-waves-in-a-periodic-potential method, are revealed [13]:

1. The extent of overlap of atomic orbitals and therefore the width of resulting energy bands depend on the spatial extension of the atomic electron wave functions. This is shown, as an example, for the case of nickel in Figure 1.3. Thus, outer s orbitals will overlap more than d orbitals and consequently s

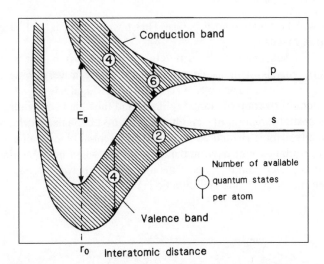

Figure 1.2.   Allowed and prohibited energy bands arising from the overlap of atomic s and p orbitals for tetrahedrally bound semiconductors like C (diamond), Si and Ge, depicted as a function of interatomic distance. At the equilibrium separation r_0 appears a forbidden energy gap E_g between the occupied and unoccupied bands arising from the sp^3 hybrid orbitals (from reference [13a], reproduced by permission of Van Nostrand).

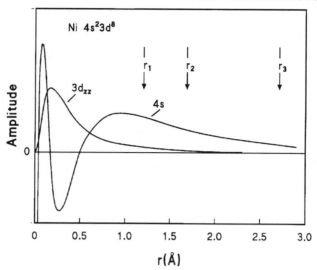

Figure 1.3. Spatial extension of the real part of the atomic s and d orbitals of solid Ni (r_1, r_2 and r_3 are first, second and third neighbour distances). There is a large overlap of atomic 4s orbital even with the second nearest neighbours. On the other hand, the 3d orbital does not extend further than the first neighbours. This situation is typical of transition metals which have a wide 4s-type band and a narrow 3d-type band (from reference [13], reproduced by permission from Springer-Verlag).

bands will be wider than d bands. This phenomenon is well documented in transition metals.

As a consequence of the different spatial extension of s and d orbitals, transition metals such as copper have typically a very wide s-type band overlapping a much narrower d-type band. (cf. Figure 1.4).

2. The extent of overlap of atomic orbitals, and thus the band width, will depend on the spatial direction of neighbouring atoms. Orbitals which are s-type are spherically symmetric and do not show preferential binding directions. But all the other orbitals will have maximum overlap when neighbouring atoms are located in preferred directions. Thus p-type orbitals will overlap preferentially in three mutually orthogonal directions:

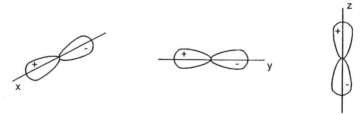

Scheme 1

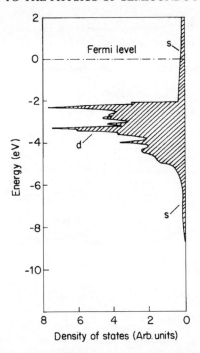

Figure 1.4. Energy band of copper due to the overlap between a narrow d band and a wide s band. There is a wide s-type band extending upward from -9 eV and a narrow d-type band between -6 and -2 eV (from reference [13], reproduced by permission from Springer-Verlag).

Figure 1.5 illustrates various possibilities of bonding of oxygen 2p orbitals with transition metal orbitals. As in the case of the hydrogen molecule, overlapping produces bonding and anti-bonding orbitals. In bonding orbitals, electrons have lower energy levels than in anti-bonding orbitals.

When there is negligible overlap, one has so-called non-bonding orbitals or states. Inside a solid, electrons occupying these non-bonding orbitals have essentially the same energies as in the free atoms.

In Figure 1.6 are shown the energy levels corresponding to bonding and anti-bonding molecular orbitals resulting from mixing of atomic orbitals of a transition metal and of six oxygens.

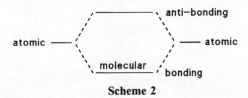

Scheme 2

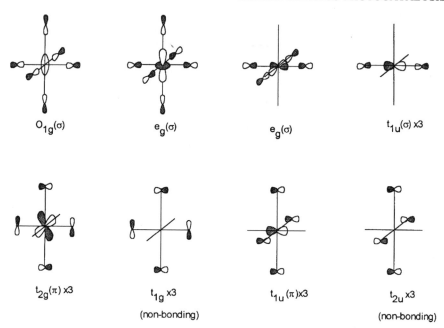

Figure 1.5. Directional mixing of transition metal 3d, 4s and 4p orbitals with the six oxygen 2p orbitals arising in an octahedron. The type of bonding (σ, π) is indicated. In the case of the triply degenerate t orbitals only one of each is drawn.

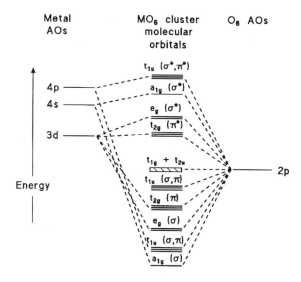

Figure 1.6. Schematic energy levels for a transition metal cation in octahedral coordination with oxygen anions.

A Mott transition [14,15] between localized and delocalized behaviour occurs when the average spatial distance between electrons exceeds about four atomic Bohr radii.

It may seem surprising that essentially similar results are obtained in an approach requiring a periodic potential as a necessary condition to describe electron behaviour inside solids, and in another which does not require such a condition. In fact, in the approach combining atomic orbitals, it is assumed that all individual anions and all individual cations have their own respectively identical average surroundings and this implies of course some kind of a periodic structure. But in any case, the attempt to correlate band structure to atomic levels is fruitful as it gives a physical basis to band width and anisotropy as a reflection of the extent and direction of overlap of atomic orbitals.

There are numerous and elaborate methods to combine the two preceding approaches, taking reasonable atomic orbitals overlapping and combining with periodically spaced neighbouring orbitals. Each of these methods (augmented plane waves, orthogonalised plane waves, Green function, pseudo-potentials) aim to be general but turn out to perform best in particular situations where the skill of the calculator for making good approximations is the determining factor. Examples of calculations and results are to be found in standard books.

5 DENSITY OF STATES IN ALLOWED BANDS

We have seen that each atomic energy level is, in fact, conserved in energy bands of solids. These atomic levels are so close together that, inside each band, they appear as a continuum. It is also found that, in every band, the number of allowed energy states per energy level is not constant. It is zero and singular (infinite derivative versus energy) at the bottom and at the top of the band. Van Hove has demonstrated that it must present also at least two other singularities inside the band as shown schematically in Figure 1.7.

Furthermore, at singularities, the group velocity of the 'electronic' travelling wave cancels:

$$\nabla_k E_k = 0$$

For the intraband singularities, the densities of allowed states are the highest and consequently it is from and to the corresponding energy levels that the maximum number of optical transitions will occur, with the additional constraint that energy and momentum conservation must be respected during such transitions.

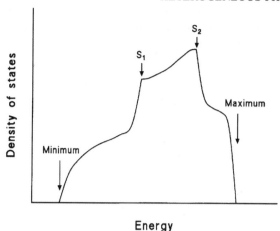

Figure 1.7. General features of the density of states function showing the different types of van Hove singularities (from reference [1], reproduced by permission from Cambridge University Press).

6   POPULATION OF ENERGY STATES: FERMI LEVEL

As is well known, electrons are fermions and, when confined inside solids, the probability of occupation of an allowed energy level, E, is

$$f(E) = \frac{1}{1 + \exp\left(\dfrac{E - E_f}{kT}\right)} \tag{1.11}$$

In this expression, E_f is the standard electrochemical potential or Fermi level which, in a first approximation, does not vary with temperature, T, and k is the Boltzmann constant.

In intrinsic semiconductors and insulators, E_f is located in an energy gap and its actual position depends on the effective masses m^* of electrons and holes at respectively the bottom and the top of the conduction and valence bands:

$$E_f = \tfrac{1}{2}E_g + \tfrac{3}{4}kT \ln \frac{m_h^*}{m_e^*} \tag{1.12}$$

Thus, in these materials the position of the Fermi level also depends on the band width because the effective masses do. This is obviously the case for oxides containing transition metal cations and, therefore, narrow d-type bands and heavy carriers.

It is well known that, in covalent semiconductors, the Fermi level is displaced by doping, that is substituting normal lattice atoms by atoms of different valency. In ionic compounds doping can also be achieved by replacing, for example, an ion of the lattice by an ion having a different atomic number (substitution as in the case of covalent semiconductors). But it can also be achieved by withdrawing or adding a normal lattice ion (non-stoichiometry). The first type of (substitution) doping can be made negligible by purification or, on the contrary, essential by controlled addition of dopants. In the same way, stoichiometry can be modified, and generally this is done, by suitable thermal treatments.

7 PHOTONS IN SOLIDS: THE ELECTRONIC DIELECTRIC FUNCTION

So far individual electrons have been considered as independent of each other and moving in an average electrical potential due to every single charged species present. In reality, direct Coulombic electron–electron interactions are strong, long range and time dependent. They cannot be easily dispensed without making the overall description irrelevant. Formally, and to include the influence of external electrical fields associated, for instance, with incident photons, this problem is treated [9,15] as a general time-dependent perturbation in the form of an electrical potential U_a applied on the one-electron states of the preceding paragraph. The general form of this perturbation of pulsation, ω, and momentum, $\mathbf{q}$, is

$$U = U_0 \exp[(i(\omega t - \mathbf{q} \cdot \mathbf{r})] \exp(i\tau t) \qquad (1.13)$$

growing exponentially from time zero with a time constant τ.

The electrical field associated with this perturbation must be negligible compared with the electric field near the nuclei, so that linear responses can be assumed. This will normally be the case (an incident flux of 1 W/cm^2 corresponds to a field of 20 V/cm, while the field around nuclei reaches 10^8 V/cm) but may not be valid with high power density laser illumination.

In ionic compounds, it is also imposed that the pulsation ω be at least one order of magnitude higher than the transverse polar optical phonons pulsations which are below about 10^{14} rad/s ($\sim 2 \times 10^{13}$ Hz).

It is then shown that, under these conditions, the actual potential acting on the electrons depends on the pulsation ω and momentum q of the perturbation:

$$U = \frac{U_a}{\varepsilon(\omega, q)} \qquad (1.14)$$

where $\varepsilon(\omega, q)$ is called the electronic dielectric function.

7.1 THE ELECTRONIC DIELECTRIC FUNCTION

In the electronic dielectric function, two physical phenomena are grouped:

1. the perturbation and response of electrons inside allowed energy bands;
2. the resonance of electrons between bands.

The most general form of ε is a complex number $\varepsilon = \varepsilon_1 + i\varepsilon_2 = 1 + \kappa_1 + i\kappa_2$ where κ is the dielectric susceptibility.

The real part of the dielectric function will contain the 'static' (frequencies very low for the electrons but nevertheless large enough so that optical phonons are not excited) electronic dielectric behaviour and the dispersive properties (index of refraction) of the solid.

On the other hand, and provided the incoming perturbation is of high enough frequency, the imaginary part of the electronic dielectric function is used to describe interband transition and therefore specifically the absorption of light.

A general expression for the electronic dielectric function in a periodic lattice is

$$\varepsilon(\omega, q) = 1 + \frac{4\pi e^2}{q^2} \sum_{k,G} \frac{[\langle k_i | \exp(i\mathbf{q} \cdot \mathbf{r}) | k_i + q + G \rangle])^2 \{f_{i(k)} - f_{f(k+q+G)}\}}{E_{f(k+q+G)} - E_{i(k)} - \hbar\omega + i\hbar\tau} \tag{1.15}$$

The sum is over all occupied and empty k states and all reciprocal lattice vectors $\mathbf{G}$; f_i and f_f are the (Fermi–Dirac) probabilities of occupation of initial and final states. The expression in the squared bracket represents the matrix element of the perturbing field. Some limiting cases are of interest:

1. when $\hbar\omega \ll E_g$ the electronic dielectric function is real and can be approximated by

$$\varepsilon = 1 + \frac{4\pi Ne^2}{m^*} \frac{\hbar^2}{E_g^2} \tag{1.16}$$

where N is the volumetric density of electrons in the highest occupied band and m^* the effective mass at the top of this band. E_g is the band gap.

This is generally written

$$\varepsilon = 1 + \left(\frac{\hbar\omega_p}{E_g}\right)^2 \quad \text{where} \quad \omega_p^2 = \frac{4\pi Ne^2}{m^*} \tag{1.17}$$

There exists a characteristic 'plasma' frequency ω_p for each partially (in metals and doped semiconductors) or completely occupied (in insulators) band. This electronic dielectric function is the static dielectric constant in the absence of 'polar' optical phonons, which is the case in covalent semiconductors like Ge and Si. The difference existing between the static $\varepsilon(0)$ and the high frequency ε (10^{16} Hz) dielectric constant (cf. Table 1.2) shows, on the one hand, the importance of polar phonons in ionic compounds and, on the other, their absence in covalent materials.

In any case, when the energy of the perturbing photon is negligible compared to E_g, the real part of the dielectric function is smaller in wide than in narrow band gap materials. $\varepsilon(0)$ depends also on the band width through the effective mass at the top of the band.

Table 1.2. Dielectric constants for some semiconductors and insulators

Materials	Static dielectric constant	Optical dielectric constant
C (diamond)	5.5	5.5
Si	11.7	11.7
Ge	15.8	15.8
InSb	17.88	15.6
InAs	14.55	12.3
InP	12.37	9.6
GaSb	15.69	14.4
GaAs	13.13	10.9
MgO	9.8	2.95
AgBr	13.1	4.6
AgCl	12.3	4
LiF	8.9	1.9

2. when $h\omega \gg E_g$, the dielectric function is again real and varies with the frequency in the same way as the dielectric function of a metal:

$$\varepsilon = 1 - \frac{\omega_p^2}{\omega^2} \tag{1.18}$$

The formula indicates that for $\omega = \omega_p$, the dielectric function is zero. This means that for this particular frequency, a small perturbation generates a large response from the electrons. This frequency is the natural plasma mode of oscillation of the electrons in the band. For Si, Ge and generally for covalent semiconductors, plasma frequencies of valence electrons lie in the range $2.5–5 \times 10^{15}$ Hz corresponding to photons of energies (10–20 eV) roughly one order of magnitude greater than the band gaps. Obviously, narrow bands, corresponding to small overlap between near neighbour orbitals have large plasma frequencies but no extensive data exist for ionic compounds.

Although only transverse electrical fields are associated with incident photons, the intraband plasma oscillations are longitudinal oscillations that can only be excited by photons of the right plasma frequency or by incident electrons. Another aspect is that for $E_g \ll h\omega < h\omega_p$, the real part of the dielectric function associated with the electrons is negative. In this range, photons are reflected by the electrons of the band. Thus, for example, Si wafers are good reflectors of visible light.

3. When $h\omega$ approaches E_g, the imaginary part $\varepsilon_2(\omega)$ of the dielectric function can no longer be neglected as photons become absorbed. $\varepsilon_2(\omega)$ can generally be expressed as a sum over occupied and unoccupied states as

$$\varepsilon_2(\omega) \div \frac{1}{\omega^2} \sum_k (\langle k_c | \Pi | k_v \rangle)^2 \cdot \delta(E_c - E_v - \hbar\omega) \tag{1.19}$$

where Π is the momentum operator. $\delta(\ )$ is the Dirac function used here to signify that the contributions to the dielectric function are zero unless the difference in energy between the initial and final states is equal to the energy of the incoming photon. Note that in the above sum $E_g(k)$ is not the minimum value of the band gap but depends on k. It is also assumed that the photon momentum is negligible so that $\mathbf{k}_c + \mathbf{k}_v = 0$. This assumption will be justified later. Assuming furthermore that the squared matrix has no significant k dependence:

$$\varepsilon_2 \div \frac{1}{\omega^2}(\langle | \ | \rangle)^2 Z_{cv}(\hbar\omega) \tag{1.20}$$

where $Z_{cv}(\hbar\omega)$ is the joint density of states.

In any case, the dielectric function has maxima (corresponding to absorption maxima) at singularities of the combined (occupied + empty) density of allowed states. These van Hove singularities occur whenever

$$\nabla_k(E_f - E_i) = 0$$

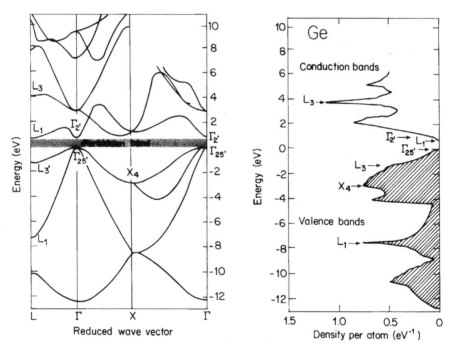

Figure 1.8. Calculated band structure $E(k)$ for Ge along directions of high symmetry (left), and the corresponding electronic density of states (right). A number of critical points (Γ,X,L), denoted according to their position in the Brillouin zone (cf. Figure 1.9) are indicated. The shaded region corresponds to the states occupied by electrons (from reference [13b], reproduced by permission from W. A. Benjamin).

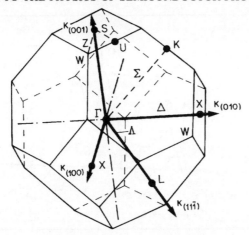

Figure 1.9. First Brillouin zone for the diamond (C, Si, Ge, GeAs...) lattice and associated designation for crystallographic directions of high symmetry. The centre of the zones is designated by Γ and energy band values are referred to the energy of this particular level. Zone boundaries in directions 100, 010, 001, etc... are designated by the letter X. Zone boundaries in all 111 directions are designated by the letter L.

Typical calculated density of states singularities at the edges and within the Brillouin zones of germanium are shown in Figures 1.8 and 1.9.

From these density of states curves the imaginary part of the electronic dielectric function can be calculated and compared to the experimental values (Figure 1.10).

As is the case with atoms, transitions are allowed when the perturbing matrix is different from zero. In such a case and provided that the bottom of the conduction band and the top of the valence band occur for the same value of k, the joint density of states and consequently the imaginary part of the dielectric function in the vicinity of the band edge is

$$\varepsilon_2(\omega) \sim Z_{ij}(\omega) \sim (\hbar\omega - E_g)^{1/2} \tag{1.21}$$

In the case of prohibited transitions near the band edge, one has the characteristic pulsation dependence:

$$\varepsilon_2(\omega) \sim (\hbar\omega - E_g)^{3/2} \tag{1.22}$$

7.2   REFLECTION, REFRACTION AND ABSORPTION

The wave vector $\mathbf{q}$ of a plane wave in vacuum is changed to $\mathbf{q}'$ after penetration inside a material. Basic Maxwell equations show that:

$$q'^2 = N^2 q^2 \quad \text{where } N^2 = \varepsilon(\omega) \text{ is the dielectric function.} \tag{1.23}$$

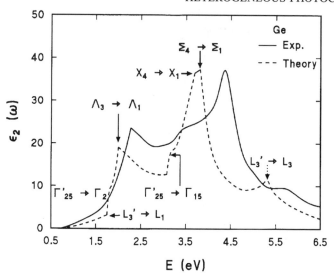

Figure 1.10. Predicted and measured variation of the imaginary part of the electronic dielectric function of germanium (from reference [9], reproduced by permission from VCH).

As we have just seen in the preceding paragraph, $\varepsilon(\omega_p)$ is generally a complex function and so is N:

$$N = n + ip$$

where n and p are real numbers. Hence

$$N^2 = n^2 - p^2 + 2inp = \varepsilon_1 + i\varepsilon_2 \tag{1.24}$$

Considering the transverse electrical field associated with an incident photon of pulsation ω and wave number q, travelling in the x-direction:

$$\mathbf{E} = \mathbf{E_0}\exp[-i(\omega t - \mathbf{q}\cdot\mathbf{x})] \tag{1.25}$$

Inside a material, this becomes

$$\mathbf{E'} = \mathbf{E_0}\exp(-i\omega t)\exp(i\mathbf{q'}\cdot\mathbf{x}) = \mathbf{E_0}\exp[-i(\omega t - n\mathbf{q}\cdot\mathbf{x})]\exp(-p\mathbf{q}\cdot\mathbf{x}) \tag{1.26}$$

which describes dispersion (dependence of the group velocity on wave number) and absorption (exponential decrease of the amplitude of the electrical field below the surface of the material). As electromagnetic energy density and flux vary with the square of the electric field, these quantities decrease below the surface as

$$\exp(-2p\mathbf{q}\cdot\mathbf{x}) = \exp(-\alpha x) \tag{1.27}$$

The coefficient α for the absorption of light energy is therefore related to both the imaginary and real part of the dielectric function. As seen above, it will have

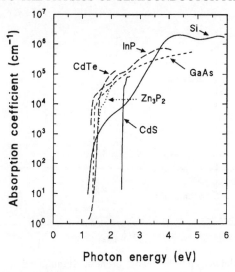

Figure 1.11. Measured absorption coefficients versus photon energy for several common semiconductors (from reference [16], reproduced by permission of John Wiley & Sons, Inc.).

characteristic variation with ω near band edges. In Figure 1.11 is represented a set of absorption coefficient's values for some selected semiconductors.

Overall, at low frequencies, electromagnetic radiations are refracted. But, whenever the real part of the dielectric function becomes negative, photons are reflected. This occurs near polar phonons frequencies and also just below the caracteristic plasma frequency of valence electrons (16 eV for Ge and Si, 12 eV for InSb). Reflections can be reduced by suitable coatings. Significant absorption occurs when the imaginary part of the dielectric function is not negligible. This occurs again at frequencies close to polar phonon frequencies and also for photon energies higher than the band-gap energy but lower than the plasma frequency.

7.3 DIRECT AND INDIRECT INTERBAND TRANSITION

Photon absorption in a material must conform locally to energy and momentum conservation laws:

$$\hbar\omega_{photon} = \text{difference in energy between initial and final state}$$

$$\hbar\mathbf{q}_{photon} = \mathbf{k_i} - \mathbf{k_f} \pm \text{crystal momentum}$$

Electron wave vectors $\mathbf{k}_i$ and $\mathbf{k}_f$ have a periodic range (period G: reciprocal lattice vector) from zero to $k_{max} = \pi/d$, where d is the near neighbour distance in a particular direction considered. The relation between electron energy and wave vector

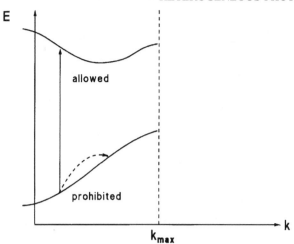

Figure 1.12. Direct (allowed) interband and (prohibited) intraband optical transitions. In a direct interband transition, energy and momentum conservation can be easily respected because photon momentum is small compared with phonon momentum. For the same reason, in a direct intraband transition, momentum conservation cannot be so easily respected.

is therefore condensed in the so-called Brillouin zone between zero and k_{max} which is of the order of $10^{10}\,m^{-1}$. On this scale, photon momentums are negligible, being of the order of $10^{7}\,m^{-1}$ in the visible. Consequently, electronic transition between occupied and empty energy states are represented vertically in $E(k)$ diagrams. Direct intraband optical transitions (cf. Figure 1.12) are obviously prohibited as, in such transitions, momentum would not be conserved.

On the other hand, forgetting the Franck–Condon principle, lattice vibrations can also participate in the conservation of momentum during photon absorption. In such a process, a phonon can be absorbed or emitted at the same time as a photon. In that case, conservation of momentum is written:

$$\mathbf{k_i} \pm \mathbf{q}_{photon} = \mathbf{k}_f \mp \Omega_{phonon} \tag{1.28}$$

where $\mathbf{k}_{i,f}$ are initial and final electron momenta and $\mathbf{q}$ and Ω, respectively, photon and phonon momenta.

Therefore, as illustrated in Figure 1.13 two characteristic types of interband transitions can occur inside materials:

1. *direct transitions*, when the participation of a phonon is not required to conserve momentum;
2. *indirect transitions* when at least one phonon participates in the absorption or emission of one photon.

It is obvious that indirect and direct transitions can occur in every material.

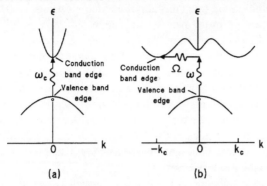

Figure 1.13. Schematic representation of a direct (a) and an indirect (b) interband transition. In both cases, photon momentum is negligible in phonon **k**-space (From reference [2]).

We have seen that the imaginary part of the dielectric function presents singularities for some frequencies which will be strongly absorbed, and these occur whenever

$$\nabla_k(E_i - E_f) = 0 \tag{1.29}$$

This condition can be fulfilled in two different ways:

1.
$$\nabla_k E_i = \nabla_k E_f = 0$$

this means that the filled band and the first empty band have, respectively, at least a maximum and a minimum for the same value of the electron wave vector. This is the ideal condition to have direct transitions.

2.
$$\nabla_k(E_i - E_f) = 0 \quad \text{but} \quad \nabla_k E_{i,j} \neq 0$$

In this case the maximum and the minimum of, respectively, the top filled band and the first empty band do not occur for the same electron wave vector. In that case, indirect transitions are favoured, at least for absorption of photons of energy corresponding to the minimum band gap.

Because an indirect transition requires the participation of a phonon of precisely the right momentum, it should occur less frequently (three-particle event) than direct transitions in direct band-gap materials. Therefore, photoexcited electrons are produced, on the average, further away from the surface in indirect- than in direct-gap materials.

Last but not least, it should be recalled that whenever a photon has the right energy to excite an electron across a band gap, the next identical photon has precisely the right frequency to stimulate the downward transition. The latter process will be most efficient in direct-gap materials. For example in a direct-gap material like GaAs the recombination rate constant of excited electrons is $10^{-12} \, m^3 \, s^{-1}$ and excited state lifetimes are in the nanosecond range.

For indirect band-gap materials, on the other hand, the absorption of a photon of minimum energy requires the participation of an adequate phonon. This necessity diminishes the chance of absorption. However, once absorption has occurred, it also takes the right kind of phonon to participate in a downward transition. Thus, in such materials, lifetimes of photo-excited electrons are longer. In indirect-gap materials such as silicon and germanium, recombination rate constant are around $10^{-17} \, m^3 \, s^{-1}$ so that the lifetimes of excited states can reach the millisecond range. As a consequence, indirect band-gap materials may also present advantages for carrying out photocatalytic processes.

However, this last conclusion is not entirely correct. Long-life excited states in indirect band-gap materials are, in fact, produced further away from the surface they should eventually reach to participate in redox reactions. And the combined effects of lifetime and distance to travel are likely to more or less compensate. To really take advantage of the longer lifetime of excited states in indirect gap material, there should be an electrical potential gradient over the zone where photons are absorbed in order to spatially separate the excited state from its ground state. It is recalled here that electrical potential gradients are present at junctions between semiconductors having different band gaps (heterojunctions), between n- and p-doped regions of the same semiconductor, and also between a metal and a semiconductor (Schottky diodes). They are also present at junctions between photocatalysts and electrolytes.

8   TRANSPORT OF EXCITED STATES

The absorption of photons in a solid perturbs the equilibrium distribution of electrons. We saw before that the amplitude of this perturbation decreases exponentially from the surface, therefore creating a concentration gradient of excited states. In the absence of any other influence, excited states diffuse down this concentration gradient away from the surface. On the other hand, if they are consumed on the surface, the concentration gradient can be bell-shaped. In the steady state, the concentration of excited states is maximum at a certain depth below the surface and there exist two flows of excited states from this maximum, one toward the surface, one away from the surface. The situation is more complicated when there is an electrical field in the region where excited states are created. Under the influence of this field, excited states of opposite charges migrate of course in opposite direction, while uncharged excited states are not moved by such a field. The isothermal variation in time of the local repartition, $F(r,k)$, in space and in momentum space of excited states obeys the general Boltzmann transport equation:

$$\frac{\partial F}{\partial t} = \mathbf{v} \cdot \nabla_{\mathbf{r}} F + \mathbf{E} \cdot \nabla_{\mathbf{v}} F + \left[\frac{\partial F}{\partial t} \right]_{scattering} \qquad (1.30)$$

The first term represents transport by diffusion in a concentration gradient, the second transport by migration in an electrical field, and the last one scattering by phonons and charged impurities. As happens often, it is easier to write a general equation than to find manageable solutions. Although the conclusions could be totally different in the case of transient phenomena, we will restrict the discussion to steady-state behaviour:

$$\frac{\partial F}{\partial t} = 0$$

Even with this restriction, the situation remains complicated because of the existence in general of neutral and charged excited states as well as of four types of phonons (transverse and longitudinal, acoustical and optical). Furthermore, transverse phonons have two degrees of freedom and materials are generally anisotropic.

In any case, considering that both migration and diffusion are limited by the same scattering process of time constant τ_i, the local current density in the steady state, for a particle of charge q_i and concentration n_i is, in a first-order approximation:

$$j_i = q_i n_i \sigma_i E + q_i D_i \, \text{grad}(n_i) \qquad (1.31)$$

where D_i is the local diffusion coefficient and σ_i is the conductivity; both are assumed to be non-tensorial (isotropic materials); E is the local total (applied + induced) electrical field.

As shown already by Einstein, D_i and σ_i are not independent:

$$D_i = \frac{\sigma_i}{n_i q_i^2} kT \qquad (1.32)$$

It is important to remember in this connection that time constants τ_i differ for electrons and for holes and may depend on position, concentration and group velocities of carriers. In any case, the absorption of photons in a solid creates a concentration gradient of electron–hole pairs below the surface. In the absence of an electric field, there will be only diffusion of electrons and holes away from the surface toward the bulk. As electrons are generally more mobile than holes, they diffuse faster thus creating an electric field accelerating the holes away from the surface toward the bulk. But, on the other hand, if holes are strongly trapped, electrons diffusing away from the surface will be pulled back by the electric field of the holes.

8.1   UNCHARGED EXCITED STATES: EXCITONS

Whenever a photon absorption is direct (does not involve the participation of a phonon) the electron and the hole created have the same momentum and therefore zero relative speed if the effective masses are not too different. Furthermore they attract each other because of their opposite charges. If they remain tied to each other by their Coulombic attraction, the resulting electrically neutral species is called an exciton. Exciton formation is evidenced by absorption of photons having energies smaller than the band gap.

In indirect band gap materials, excitons cannot be produced for energies corresponding to the minimum band gap as the electron and hole created by light absorption have different momenta. Furthermore, any exciton produced by a direct transition in these materials may decay to a free electron and a free hole having different momenta.

Stable excitons are therefore most likely to be found in direct-gap materials. In one type of approximation, an exciton being a bound electron–hole pair is treated as a hydrogen atom where the electron is bound to a proton by a spherically symmetrical Coulombic potential. In a solid though, this Coulombic potential is reduced by the appropriate dielectric constant and the masses of the interacting particles are effective masses. The allowed energies for such an atom are quantified:

$$W_n = \frac{m^*e^4}{2\varepsilon^2\hbar^2}\frac{1}{n^2} \tag{1.33}$$

where m^* is the effective reduced mass of the coupled electron and hole, n is an integer and ε is the dielectric function of the material for $\omega = W_n/h$. It is assumed, as a first approximation, that the allowed energy levels for the excited electron are located at W_{in} below the conduction band.

This kind of 'hydrogenic' exciton (Wannier–Mott) seems to represent the situation in materials with light carriers rather well (small band gap, large s or σs type bands). But for most solids and especially so for ionic compounds, ε is large, so that these types of excitons lie so close to the bottom of the conduction band that they are thermally dissociated at room temperature. For instance, as shown in Table 1.3, in GaAs, the exciton binding energy is around 0.004 eV, that is 1/7 of kT at 300 K.

In ionic materials there also exists a second type of exciton (Frenkel) corresponding to excited states of the anions. They are generally localised on anion sites and their binding energies may reach 1 eV.

It should be emphasised that excitons are neutral species that do not contribute to the conductivity. They do not migrate in an electrical field. On the other hand, they diffuse down concentration gradients.

Table 1.3. Binding energies of excitons for some selected materials in meV (1 meV = 10^{-3} eV)

Si	14.7
Ge	4.1
GaAs	4.2
GaP	3.5
InP	4.0
CdS	29
CdSe	15
BaO	56
KCl	400
KI	480
RbCl	440
AgCl	30
TlCl	11

8.2 CHARGED EXCITED STATES: POLARONS

The accelerated motion of charged species in electrical fields is perturbed by longitudinal phonons in two different ways. First of all, for carriers having kinetic energies that are small compared to their band width, longitudinal phonons, whether they are acoustical or optical, are equivalent to mechanical strains applied to the materials. They modify the average distance between nuclei and, chaotically, the periodicity of the average electrical potential acting on the carriers [17]. Transverse phonons, which do not produce first-order changes in volume, do not contribute to this type of scattering.

In the case of such quasi-elastic scattering by longitudinal phonons, a mean free path between scattering events can be introduced:

$$l \sim (m^*T)^{-1}$$

The mean free path, l, and therefore the mobility, decrease with an increase of the temperature but also with increasing effective mass and therefore with decreasing band width and increasing band gap.

For Si at room temperature the mean free path is 2–3 um. However, in ionic compounds charged excited species also interact with longitudinal optical phonons via their electrical field: electrons attract cations and holes attract anions. Both types of charged carriers are therefore surrounded by a sort of cloud of optical phonons.

A carrier with its associated optical 'polar' phonon cloud is called a polaron. The effective mass m^*_{pol} is greater than the bare band mass m^* of the carrier and the increase in mass is a measure of the number of phonons associated with each polaron. In Table 1.4, this number is 0.5 α.

Table 1.4. Polaron coupling constants α, masses m^*_{pol}, and band masses m^* for electrons in the conduction band of various crystalline materials

	KCl	KBr	AgCl	AgBr	ZnO	PbS	InSb	GaAs
α	3.97	3.52	2.00	1.69	0.85	0.16	0.014	0.06
m^*_{pol}/m	1.25	0.93	0.51	0.33	—	—	0.014	—
m^*/m	0.50	0.43	0.35	0.24	—	—	0.014	—
m^*_{pol}/m^*	2.5	2.2	1.5	1.4	—	—	1.0	—

As long as the number of phonons attached to each carrier is not too high, polarons are considered to move in a polaron band and they are scattered by longitudinal acoustical and optical phonons with a mean free path as above. The mobility of these so-called 'large' polarons decreases with increasing temperature and so does the associated conductivity. When the number of phonons associated with the charged excited species becomes sufficiently high, they become so heavy that they must be considered as trapped. They are then called small polarons. The trapping, however, is not perfect and small polarons can move from one trapped position to another by two different mechanisms:

1. At low temperatures, roughly below half the Debye temperature, carriers can tunnel between sites and the tunneling rate decreases when the distance between sites increases as the temperature is raised.
2. At higher temperature, carriers can 'hop' through the phonon cocoon that traps them and reach another position where they wrap themselves up again in polar phonons. In that case the hopping rate, and thus the mobility, increases when the temperature rises.

9 EXAMPLES

From the known types of atomic orbitals that combine to insure cohesion of condensed materials, and from crystallographic data, certain properties of their electronic band structures can be inferred. The influence of lattice distortion and of non-stoichiometry can also, qualitatively, be discussed.

9.1 COVALENT SEMICONDUCTORS (Si, Ge, III–V COMPOUNDS)

These materials have a cubic diamond structure in which each atom is surrounded by four isoelectronic neighbours. For each lattice site, the one s and three p atomic orbitals form four identical hybrid sp^3 orbitals reaching toward the four corners of a tetrahedron. The valence and conduction band are formed by overlap of these hybrid orbitals. The amount of overlap is large as they have both an s-like and a directional p character.

Band widths are in the range 15–20 eV. Band gaps are between 0.2 and 2 eV. Therefore effective masses near band edges are between 1/10 and 1/100 of the free electron mass. The carrier mobilities vary accordingly (Table 1.1).

9.2 INFLUENCE OF THE TYPE OF OVERLAPPING ORBITALS: SnO_2 AND TiO_2

In some compounds, bonds and bands are formed by overlapping of orbitals of different character. For instance in oxides, cation and oxygen ion s-type orbitals have large spatial extension and will overlap strongly giving wide s-type energy bands. Large overlap also occurs with p-type orbitals pointing toward each other. The bands are then called pσ.

On the other hand, p-type orbitals of the cation and of its surrounding oxygen anions which are parallel to each other will overlap less, forming narrower pπ bands. Even narrower bands will be obtained from d-type orbitals overlapping with pπ oxygen orbitals.

Non-overlapping orbitals form non-bonding states.

Given octahedral or tetrahedral coordination around a cation having known s, p, d, ... orbitals, qualitative band structures can be suggested respecting the chemical stoichiometry. In that respect, a comparison between two oxides of the same stoichiometry and crystallographic structure illustrates the method. Two such oxides are SnO_2 and TiO_2 (rutile). For these compounds [12], oxygen anions form hexagonal close packed layers, and tin and titanium cations are located in octahedral sites to obtain the right stoichiometry. Each oxygen has three coplanar near neighbour cations at 120° to each other.

Band gaps are of the same order of magnitude:

SnO_2 : 3.59 eV direct forbidden TiO_2 (rutile) : 3.03 eV direct forbidden

and so are lattice parameters:

$$SnO_2 \quad a = b = 0.437 \quad c = 0.3285 \quad (nm)$$

$$TiO_2 \quad a = b = 0.459 \quad c = 0.296 \quad (nm)$$

But in tin oxide, the electronic bands are due to the overlap of s and p orbitals. In fact the valence band is mainly composed of O_{2p} orbitals while the conduction band consists mainly of Sn_{5s} and Sn_{5p} orbitals. Neither band is as wide as in covalent semiconductors, being about 4 eV for the conduction and 10 eV for the valence band (Figure 1.14 a,b).

The effective masses of conduction electrons are $0.299m_o$ (perpendicular to c) and $0.234m_o$ (//c) comparable to the value in Si ($m^* = 0.33m_o$). However, their mobility (2.4×10^{-2} m²/(V s) at 300 K is not as high (Si: 1.6×10^{-1}). The static electronic dielectric constant is 14, decreasing to 4 at high frequency.

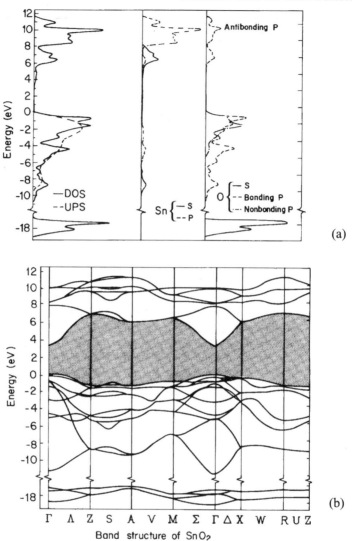

Figure 1.14. (a) Density of states (DOS) and contribution of s and p orbitals to the SnO_2 energy bands. DOS is the calculated density of states and UPS the measured intensity of the ultraviolet photoelectron emission. (b) Band structure of SnO_2 in the main crystallographic directions (From reference [12], reproduced by permission from the Institute of Physics).

In titanium oxide (rutile) the conduction and to some extent also the valence band have essentially a d-like character and are therefore much narrower than in SnO_2. As a consequence, the effective mass of carriers in the conduction band is between 10 and $20m_o$ and the Hall mobility at room temperature is reduced to around $4 \times 10^{-5} \, m^2/(V \, s)$. Such values correspond to small polarons. Further evidence for the existence of small polarons in TiO_2 is the variation of mobility with temperature switching from a tunnelling mechanism at low temperature to a thermally activated hopping at high temperature.

In rutile, the static dielectric electronic constant is around 90–100 decreasing to 6–7 at visible light frequencies.

The elastic constants of both solids being of comparable magnitude, acoustic phonons are not considered to be responsible for the difference in carrier mobilities and masses. In the same way, longitudinal optical phonons are not considered for the difference in masses as their coupling to electrons would produce polarons of essentially the same size in both materials.

So, if polarons are large (small number of phonons per carriers) in SnO_2, and small (self trapped) in TiO_2, it is because of the different band widths: in tin oxide, the conduction band is large because of strong s and p character, while in titanium oxide, the conduction band is narrow because of its d character.

9.3 LATTICE DISTORTION AND DEVIATION FROM STOICHIOMETRY: SiO_2

In SiO_2, each Si has four hybrid sp^3 orbitals pointing toward the corners of a tetrahedron of oxygen ions, and each oxygen ion forms bonds to two Si cations.

The oxygen s-type state is so low in energy compared to the other states that it remains atom-like, forming a deep, wide and completely filled band. As shown in Figure 1.15, the hybrid sp^3 orbitals of silicon couple to the directional, say p_x and p_y, orbitals of oxygen. The remaining p_z orbital of oxygen is non-bonding.

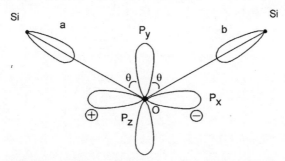

Figure 1.15. Si–O–Si unit for building the molecular model with the two sp^3 hybrid orbitals of the Si atoms and the three p orbitals of one oxygen atom. The p_z orbital is perpendicular to the plane of the figure (from reference [9], reproduced by permission from VCH).

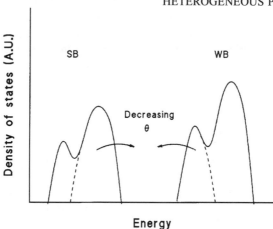

Figure 1.16. Schematic variation with Si–O–Si bond angle θ, of the energies of strong bonding (SB) and weak bonding (WB) band of SiO_2 (from reference [9], reproduced by permission from VCH).

The amount of overlap between the Si (sp^3) and the O (p_x and p_y) orbitals, and therefore the width and shape of the obtained band, depends on the Si–O–Si angle. Schematically, one expects that the valence band is made of two sub-bands, one associated with overlap with oxygen p_x-like orbitals (strong bonding) and the other with p_y-like orbitals (weak bonding). These two sub-bands differ in energy by an

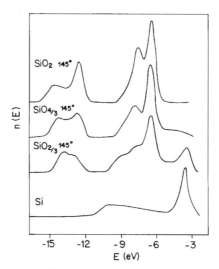

Figure 1.17. Theoretical density of valence band states for SiO_x ($x = 2$, 4/3, 2/3, 0) (from reference [9], reproduced by permission from VCH).

amount which depends on the Si–O–Si bond angle [9]. Thus, in a traditional density of states versus energy diagram, when the angle decreases, the 'strong bonding' band rises while the 'weak bonding' band sinks (Figure 1.16).

The influence of non-stoichiometry can also be qualitatively predicted. For example, as shown in Figure 1.17, when O ions are missing in the lattice, unbounded atomic Si states are introduced at an energy level corresponding to the free atom. If more and more oxygen ions are missing, these Si levels can overlap to form a new Si-type band (Figure 1. 17).

There is evidence for a similar behaviour in understoichiometric 'reduced' TiO_2.

9.4   INFLUENCE OF CRYSTALLOGRAPHIC STRUCTURE: RUTILE VERSUS ANATASE

Titanium dioxide is used as a photocatalyst in two crystallographic structures: rutile and anatase. In both structures, each titanium ion is at the centre of an oxygen octahedron. In rutile, the oxygen ions form a slightly distorted hexagonal compact lattice. In anatase, the oxygens form a cfc lattice.

In both structures, each oxygen has three coplanar near neighbour titanium cations.

But in rutile, the three Ti–O–Ti angles are roughly equal to 120°. In anatase, on the other hand, one Ti–O–Ti angle is about 180° while the two other are close to 90°. If the same conclusions as were drawn in the case of the varying Si–O–Si

Table 1.5.   Selected physical properties of rutile and anatase TiO_2

	Rutile	Anatase
Band-gap energy	3.03 eV	3.2 eV
Electron effective mass	$m^* = 20m_o$	$m^* = 1m_o$
Hall mobility	0.1 cm²/(V s)	4 cm²/(V s)
(electron, room temperature)		
Dielectric constant		
$\varepsilon_s(\perp c)$	89	31
$\varepsilon_s(\| \|c)$	173	170

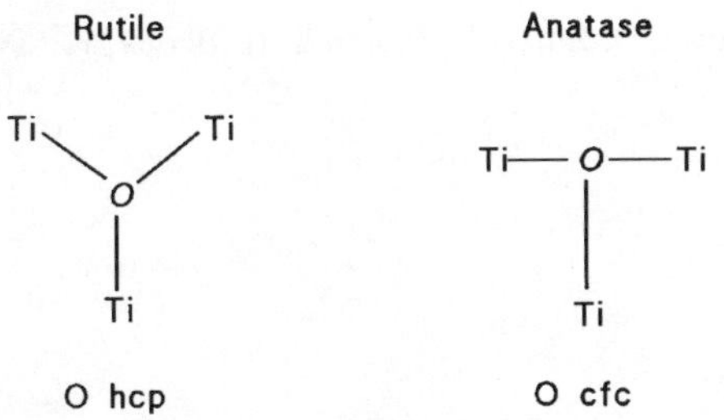

Rutile   Anatase

O hcp   O cfc

Scheme 3

angle can be extended to the anatase case, their result will be a significant widening of the d bands with an accompanying decrease in effective masses and an increase in mobility.

REFERENCES

[1]　J. M. Ziman, *Principles of the Theory of Solids*, Cambridge University Press (1972)

[2]　Ch. Kittel, Introduction to Solid State Physics, Wiley, New York (1986)

[3]　O. Madelung, *Introduction To Solid State Theory*, Springer-Verlag, Berlin (1978)

[4]　M. H. Cohen, Introductory Lectures, The Optical Properties of Solids, Rendiconti S:I:F:XXXIV

[5]　Y. Ayant, E. Belorizky, *Cours de Mécanique Quantique*, Dunod, Paris (1974)

[6]　M. Born, R. Oppenheimer, *Ann. Phys.* (Leipzig) **84**, 457 (1927)

[7]　L. Brillouin, *Wave Propagation in Periodic Structures*, Academic Press, New York (1960)

[8]　M. H. Cohen, H. Fritzsche, S. R. Ovshinsky, *Phys. Rev. Lett.*, **22**, 1065 (1969)

[9]　M. Lannoo, in *Materials Science and Technology*, Vol. 4, *Electronic Structure and Properties of Semiconductors*, ed. W. Schröter, VCH, Weinheim (1991)

[10]　J. B. Goodenough, *Les Oxydes des Métaux de Transition*, Gauthier-Villars, Paris (1973)

[11]　J. R. Chelikowsky, A. Franciosi, eds., *Electronic Materials*, Springer-Verlag, Berlin (1991)

[12]　N. Tsuda, K. Nasu, A. Yanase, K. Siratori, *Electronic Conduction in Oxides*, Springer-Verlag, Berlin (1991)

[12a]　J. Robertson, *J. Phys. C: Solid State Phys.*, **12**, 4767 (1979)

[13]　H. Ibach, H. Lüth, *Solid State Physics*, Springer-Verlag, Berlin (1991)

[13a]　M. Shockley, *Electrons and Holes in Semiconductors*, Van Nostrand, New York (1950)

[13b]　F. Herman, R. L. Kortum, C. D. Kuglin, J. L. Shay, in *II–VI Semiconducting Compounds*, ed. D. G. Thomas, W. A. Benjamin, New York (1967)

[14]　N. F. Mott, *Can. J. Phys.* **34**, 1356 (1958)

[15]　E. Wigner, *Trans. Faraday Soc.* **34**, 678 (1938)

[16]　A. L. Fahrenbruch, R. H. Bube, *Fundamentals of Solar Cells*, Academic Press, New York (1981)

[17]　J. R. Bardeen, W. Shockley, *Phys. Rev.*, **80**, 72 (1950)

[18]　H. Tang, K. Prasad, R. Sanjinès, P. E. Schmid, F. Lévy, *J. Appl. Phys.* **75**, 2042 (1994)

[19]　R. Sanjinès, H. Tang, H. Berger, F. Gozzo, G. Margaritondo, F. Lévy, *J. Appl. Phys.* **75**, 2945 (1994)

[20]　L. Forro, O. Chauvet, D. Emin, L. Zuppiroli, H. Berger, F. Lévy, *J. Appl. Phys.* **75**, 633 (1994)

2 Surface versus Bulk Properties

A. ATREI and G. ROVIDA
Dipartimento di Chimica, Università di Firenze, 50129 Firenze, Italy

1 INTRODUCTION

The surface properties of a solid are largely determined by the different environment experienced by the atoms at the surface in comparison with those in the bulk. Surface atoms have fewer neighbours, thus they are more weakly bonded. This implies that work is needed to create a new surface, since interatomic bonds must

Heterogeneous Photocatalysis, Edited by M. Schiavello
© 1997 John Wiley & Sons Ltd.

be broken; this surface energy depends on the atomic structure of the surface and will thus vary for different crystal faces of the solid. The change in the bonding at the surface is also responsible for the observed surface relaxation and reconstruction, when atoms at the surface move to new positions, different from those expected for a simple truncation of the infinite crystal.

The knowledge of the atomic structure at the surface is fundamental in order to understand the surface properties of a solid and to base an interpretation of any surface process. Thus, the most important step in surface characterisation is the determination of the structure at an atomic level.

The change in the electron distribution around the surface atoms causes a surface dipole moment and this contributes to the energy needed to remove an electron from the solid, the work function. The presence of a surface also causes the appearance of new electron wave functions, the so-called surface states, localised at the surface. These states affect the electronic properties of the surface and may play an important role in the interaction with adsorbed atoms or molecules.

The different environment at the surface also changes the atomic vibrations of surface atoms. Moreover, the mobility of atoms at the surface is much higher than in the bulk because of the lower energy required to migrate from a site to another. Thus, surface diffusion can be important at temperatures much lower than those required for bulk diffusion.

For a multicomponent solid, the composition at the surface can differ significantly from that of the bulk. The exact surface composition cannot easily be derived theoretically from that of the bulk, so that it must be determined with suitable experimental techniques. In some cases, the segregation of one component at the surface can strongly affect the physical and chemical properties.

The aspects mentioned above will be treated in this chapter at an introductory level.

Our knowledge of surface structure and properties has greatly increased in the last 30 years, and is to be attributed to the development of powerful experimental techniques which allowed us to derive direct information on the atomic structure, and composition of the first few atomic layers. Nowadays, the number of surface techniques has increased so much that a list of the most common acronyms used to indicate them is quite long, as reported in Table 2.1. Only a few of them will be briefly described in this chapter, in order to give the reader an idea of the applicability of the most important techniques to the study of solid surfaces.

2 SURFACE STRUCTURE

If we consider the surface of a powdered oxide or of a polycrystalline metal sheet, the surface structure of this material cannot be unequivocally determined on an atomic scale, since it will strongly vary depending on the site examined. But if we cut a single crystal of a material along a given crystallographic plane and we polish

Table 2.1. Techniques for the study of solid surfaces

Acronym	Technique	Composition	Structure	Electronic properties (chem. state)	Vibrational properties
AES	Auger electron spectroscopy	X		(X)	
AFM	Atomic force microscopy		X		
APS	Appearance potential spectroscopy	X		X	
ESCA	Electron spectroscopy	X		X	
EELS	Electron energy loss spectroscopy			X	
ESD	Electron stimulated desorption	X		X	
ESDIAD	ESD ion angular distribution		X	X	
FEM	Field emission microscopy			X	
FIM	Field ion microscopy		X		
HEIS	High-energy ion scattering	X	X		(X)
HREELS	High-resolution electron energy loss spectroscopy	(X)	(X)	(X)	X
INS	Ion neutralisation spectroscopy			X	
IPS	Inverse photoemission spectroscopy			X	
IRAS	Infrared reflection absorption spectroscopy				X
LEED	Low-energy electron diffraction		X		(X)
LEIS	Low-energy ion scattering	X	X		
MBS	Molecular beam scattering		X		
PSD	Photon stimulated desorption	X		(X)	
RHEED	Reflection high-energy electron diffraction		X		
SEXAFS	Surface extended X-ray absorption fine structure		X		
SIMS	Secondary ion mass spectroscopy	X			
STM	Scanning tunnelling microscopy		X		
SXRD	Surface X-ray diffraction		X		(X)
TDS	Thermal desorption spectroscopy (see TPD)				
TPD	Temperature programmed desorption	X			
UPS	Ultraviolet photoelectron spectroscopy		(X)	X	
WF	Work function measurement			X	
XPD	X-ray photoelectron diffraction		X		
XPS	X-ray photoelectron spectroscopy	X		X	

the surface as finely as possible, we can expect that, after a proper treatment of cleaning and annealing in a controlled environment in order to remove the surface impurities and the structural damage caused by the cutting process, most of the atoms at the surface will be found in large flat domains with the same perfectly periodic structure, while a minority will be found at steps, kinks and other structural defects. This is indeed what has been found for a large variety of single crystal faces. The problem is to determine the exact atomic composition and structure in relation to those of the bulk crystal, which are usually well known. Since, however, the number of atoms present in the first atomic planes is very low in comparison to that of atoms in the bulk, this kind of study is not so easy.

It is the availability of high-purity single crystals of most solids that has allowed, in the last thirty years, the development of several experimental techniques which can give direct information on the atomic structure of solid surfaces [1]. Among these, the first to be applied was low- energy electron diffraction (LEED), which is still the most widely used for its relatively low cost and experimental simplicity. Since the early 1960s (when the progress in technology made LEED experiments much easier in comparison with the first one of Davisson and Germer in 1927) its use has rapidly increased, giving rise to a new field of research, which can be defined as 'surface crystallography'. In the last ten years, however, several new techniques, such as X-ray photoelectron diffraction (XPD), scanning tunneling microscopy (STM) and X-ray surface diffraction, have been developed, particularly in the study of systems that cannot be successfully investigated by LEED alone. Thus, for example, XPD can be applied to study the structure of the first stages of the epitaxial growth of a new surface phase, STM allows the atomic structure of surface defects and their movements to be directly observed, while X-ray surface diffraction (as recently performed at very low incidence angles and using intense synchrotron radiation) opened the possibility of determining the atomic positions in the case of complex structures.

2.1 SURFACE CRYSTALLOGRAPHY

The 'surface crystallography' of a solid can be considered the structure of the first atomic layers at and below the 'geometrical surface' to a depth such that the positions of the atoms do not differ appreciably from those in the bulk. This depth cannot be evaluated *a priori*, since it may vary depending on the solid considered. However, in the case of metals, significant structural differences do not usually extend more than two or three atomic layers inside the solid, while for semi-conductors like Si, Ge, GaAs, an appreciable perturbation can be found up to the fourth or fifth layer.

We can imagine an 'ideal surface' obtained by cutting an infinite crystal into two semi-infinite parts along a given plane and assuming that the interatomic distances in the region close to the surface do not change. The atoms will be arranged in a lattice which is perfectly periodic only in the two directions lying in the surface plane, the periodicity in the third dimension having been interrupted by the surface plane. For atomic planes parallel to the surface we can thus define two vectors **a** and **b** such that any vector $\mathbf{t} = m\mathbf{a} + n\mathbf{b}$ (with m and n integers) corresponds to a translation connecting two equivalent atom positions in the same plane. The two vectors define the two-dimensional (2D) unit cell (or unit mesh) which will correspond to one of the five 2D Bravais lattices, as shown in Figure 2.1.

In each atomic plane parallel to the surface, a given family of atomic rows is identified by two Miller indices (hk), which are the 2D equivalent of the Miller indices (hkl) characterising a given family of atomic planes in 3D crystallography. Inside the unit mesh, the atoms can be arranged according to the 17 plane groups which can be obtained combining translation with the symmetry elements operating in a plane (rotation axes and symmetry planes perpendicular to the surface).

In the case of the 'ideal surface' considered, all the atomic planes parallel to the surface would have the same periodicity (unit mesh) and the same interplanar spacing as in the bulk. In the case of a real surface, atoms near the surface will generally relax to new equilibrium positions, so that we must expect (as verified in some cases) both a change in the spacing of the outermost planes and a change of the periodicity along the surface plane (i.e., variation of the unit mesh): in the latter case we speak of a true 'reconstruction' of the surface structure.

For stronger reasons, such changes are to be expected in the case of adsorption of atoms or molecules on the surface.

In both cases of a reconstructed clean surface and of a surface with the presence of an adsorbate, the new surface structure will correspond to a new unit mesh with

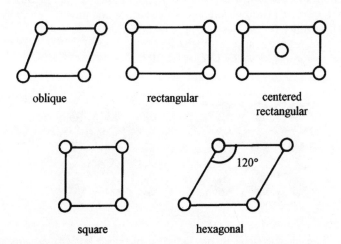

Figure 2.1.   The five two-dimensional Bravais lattices.

primitive vectors $\mathbf{a}_S$ and $\mathbf{b}_S$, which can be expressed in terms of the ideal or bulk mesh:

$$\mathbf{a}_S = g_{11}\mathbf{a} + g_{12}\mathbf{b}$$
$$\mathbf{b}_S = g_{21}\mathbf{a} + g_{22}\mathbf{b}$$

where g_{ij} are the elements of a matrix G which can be used to identify the surface unit mesh. For example, the structure in Figure 2.2(a), which can be assumed to derive from adsorption of oxygen atoms on a square array of atoms, like the (100) plane of a face-centred cubic (fcc) metal, can be indicated as

$$\begin{pmatrix} 2 & 0 \\ 0 & 2 \end{pmatrix}$$

while that in Figure 2.2(b) corresponds to

$$\begin{pmatrix} 1 & 1 \\ 1 & -1 \end{pmatrix}$$

This notation is the most complete and can be used in all cases. Moreover, the determinant of G directly gives the ratio between the area of the surface mesh and that of the ideal mesh. This allows the surface coverage to be derived immediately, in the case of an adsorption structure, if the number of adsorbed atoms per unit mesh is known.

In a more compact notation, the four elements of the G matrix are written on the same line: $(g_{11}, g_{12}, g_{21}, g_{22})$. Thus, the two above structures can be written (1 1, −1 1) and (2 0, 0 2), respectively.

Another kind of notation, which is of more common use, can be applied when the angles between the primitive vectors of the surface and the bulk unit meshes are the same [2]. In this case the new mesh is identified by

$$\left({}^{a_s}/_a \times {}^{b_s}/_b \right) R\alpha$$

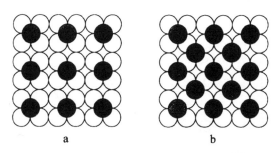

Figure 2.2. Two simple models of surface structures: (a) (2×2) structure; (b) $(\sqrt{2} \times \sqrt{2})R45°$ structure.

where α is the angle of rotation between the two meshes (omitted if zero), in the case of the structures shown in Figure 2.2(a, b), they can be indicated as (2×2) and $(\sqrt{2} \times \sqrt{2})R45°$, respectively. In some cases a centred unit mesh is chosen for simplicity and the letter c is placed before the parentheses: thus, the structure of Figure 2.2(b) can also be indicated as $c(2 \times 2)$, while that of Figure 2.2(a) is better defined as $p(2 \times 2)$ (i.e., primitive unit mesh). In the case of structures formed by adsorbates, the symbol of the atom or molecule is placed after the expression. For instance, if the above structures correspond to adsorption of oxygen atoms on the (100) face of nickel, they can be respectively indicated as $Ni(100)$-$c(2 \times 2)$-O or $Ni(100)$-$(\sqrt{2} \times \sqrt{2})R45°$-O, and $Ni(100)$-$p(2 \times 2)$-O.

As in the case of 3D structures, for 2D structures as well the concept of reciprocal lattice is used to describe the diffraction mechanism better, as will be shown later. The 2D reciprocal lattice is formed by an array of points which is periodic in the two dimensions parallel to the surface plane; these points, with respect to one of them chosen as the origin, define vectors

$$\mathbf{q}_{hk} = h\mathbf{a}^* + k\mathbf{b}^*$$

where h and k coincide with the Miller indices of the atomic rows perpendicular to the vector $\mathbf{q}_{hk}$ in the direct lattice; $\mathbf{a}^*$ and $\mathbf{b}^*$ are the primitive vectors identifying the reciprocal unit mesh. These vectors are related to the primitive vectors of the direct lattice by the relations

$$\mathbf{a} \cdot \mathbf{b}^* = \mathbf{b} \cdot \mathbf{a}^* = 0$$
$$\mathbf{a} \cdot \mathbf{a}^* = \mathbf{b} \cdot \mathbf{b}^* = 2\pi$$

$$a^* = \frac{2\pi b}{|\mathbf{a} \times \mathbf{b}|} \qquad b^* = \frac{2\pi a}{|\mathbf{a} \times \mathbf{b}|}$$

that is, $\mathbf{a}^*$ has a modulus which is inversely proportional to the distance of the rows parallel to $\mathbf{b}$ and is perpendicular to $\mathbf{b}$, the same for $\mathbf{b}^*$ in relation to $\mathbf{a}$.

It can be shown that each reciprocal lattice vector $\mathbf{q}_{hk}$ is not only perpendicular to the atomic rows defined by the two Miller indices (h, k), but its modulus is given by

$$q_{hk} = \frac{2\pi}{d_{hk}}$$

where d_{hk} is the distance of (hk) atomic rows.

In the case of a surface structure with periodicity differing from that of the bulk, the surface reciprocal lattice can be described in terms of the reciprocal lattice corresponding to the 2D periodicity of the parallel planes in the bulk. Indicating with $\mathbf{a}_S^*$ and $\mathbf{b}_s^*$ the primitive vectors of the surface reciprocal lattice, we have

$$\mathbf{a}_s^* = g_{11}^* \mathbf{a}^* + g_{12}^* \mathbf{b}^*$$
$$\mathbf{b}_s^* = g_{21}^* \mathbf{a}^* + g_{22}^* \mathbf{b}^*$$

where g_{ij}^* are the elements of matrix G^*. It can be shown that the matrix G^* is the inverse transposed of matrix G of the direct lattice.

2.2 TECHNIQUES FOR SURFACE STRUCTURE DETERMINATION

2.2.1 Low-energy Electron Diffraction (LEED)

LEED is the most widely used experimental technique for the study of the structure of solid surfaces. We can say that surface crystallography is related to LEED in the same way as classical crystallography is related to X-ray diffraction.

In LEED technique, a collimated beam of monoenergetic electrons is used with energy usually in the range 30–300 eV. The sample on which the electrons impinge must be a single crystal cut along the face to be studied. The crystal must have a diameter of several millimetres, since the electron beam diameter is of the order of 1 mm. When the beam strikes the surface plane, only a small fraction of the electrons is backscattered elastically. Most of them suffer a loss of energy due to inelastic events, producing the emission of new electrons, the so-called secondary electrons. In LEED, only the angular distribution of the electrons scattered elastically is studied, so that all the other electrons must be removed before they reach the detector. To detect the scattered electrons, a fluorescent screen is used in most systems, so that the diffracted beams give rise to spots on the screen. Therefore, the entire diffraction pattern can be directly observed through a window of the vacuum chamber. The intensity of each diffraction spot can be measured using a spot photometer or a video camera.

In a display type LEED apparatus (Figure 2.3) the electrons coming from the sample surface are first filtered in order to eliminate all the electrons with energy different from that of the primary beam. This is accomplished by a set of at least two grids, the first being held at the same potential of the sample, the second at a potential close to that of the emitting cathode: the latter grid has the function of repelling the inelastic electrons. The elastically scattered electrons are then accelerated by a strong (several kilovolts) potential onto the fluorescent screen. The grids and the screen are usually of a spherical shape with the sample in the geometrical centre in order to avoid any distortion of the electron trajectories.

Of course, the LEED optics must be placed in an apparatus in which ultra-high vacuum conditions (10^{-9}–10^{-10} torr) can be provided, as for all other techniques for surface studies, in order to prepare and maintain a clean surface.

The sample surface is cleaned *in situ*, in general by repeated cycles of ion bombardments and annealings at high temperatures. The thermal treatments are needed in order to restore the long-range order in the surface atomic planes damaged by the ion impact. In a few cases, a clean surface can be produced by cleavage under vacuum. When feasible, this method provides the most perfect surfaces.

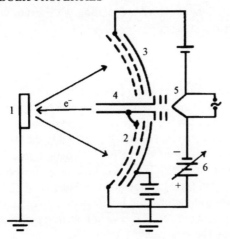

Figure 2.3. Schematic LEED apparatus: (1) sample; (2) grids; (3) fluorescent screen; (4) electron gun; (5) cathode; (6) electron beam voltage.

The diffraction mechanism can be easily understood considering the primary beam as a plane wave with wave function

$$\psi_i = A \exp(i\mathbf{k}^0 \cdot \mathbf{r}) \tag{2.1}$$

where $|\mathbf{k}^0|^2 = 2E$ (using atomic units, so that $(h/2\pi)^2 = m = e = 1$), while the scattered electron's wave function, due to the lattice periodicity along the two dimensions parallel to the surface plane, can be written in the Bloch form

$$\psi_S = \exp(i\mathbf{k}^0_{//} \cdot \mathbf{r}_{//}) \cdot u_S(\mathbf{r}) \tag{2.2}$$

where $\mathbf{k}^0_{//}$ and $\mathbf{r}_{//}$ are the components of $\mathbf{k}^0$ and $\mathbf{r}$ parallel to the surface, while $u_S(\mathbf{r})$ is a function having the same periodicity of the surface plane and can be written as a Fourier expansion:

$$u_S(\mathbf{r}) = \sum_{\mathbf{q}} \alpha_{\mathbf{q}}(z) \exp(i\mathbf{q} \cdot \mathbf{r}_{//}) \tag{2.3}$$

where $\mathbf{q}$ are the vectors of the 2D reciprocal lattice (identified by the two indices (hk), see above); the coefficients $\alpha_{\mathbf{q}}(z)$ are a function of the coordinate z normal to the surface. Substituting equation (2.3) in the expression of ψ_S (equation (2.2) and resolving the Schrödinger equation, it can be shown [3,4] that the coefficients $\alpha_{\mathbf{q}}(z)$ have the form

$$\alpha_{\mathbf{q}}(z) = A_{\mathbf{q}} \exp\left[\pm i \left(2E - |\mathbf{k}^0_{//} + \mathbf{q}|^2\right)^{1/2} z\right] \tag{2.4}$$

Choosing the minus sign, since we are dealing with waves coming back from the surface, we find for ψ_S

$$\psi_S = \sum_{\mathbf{q}} A_{\mathbf{q}} \exp\left\{ i\left[(\mathbf{k}_{//}^0 + \mathbf{q}) \cdot \mathbf{r}_{//} - (2E - |\mathbf{k}_{//}^0 + \mathbf{q}|^2)^{1/2} z \right] \right\} \qquad (2.5)$$

We can seen that the scattered wave is formed by a series of plane waves, i.e., the diffracted beams: each beam corresponds to a vector of the 2D reciprocal lattice with indices (hk) and has a wave vector component on the surface given by $\mathbf{k}_{//}^0 + \mathbf{q}$. Thus, the angles of the diffracted beams are related to the size and symmetry of the unit mesh, but the geometry of the diffraction pattern does not give information about the positions of the atoms inside the unit mesh. This piece of information is contained in the amplitudes $A_{\mathbf{q}}$ of the scattered beams.

From the above considerations, the LEED pattern is composed of a set of beams, each corresponding to a vector $\mathbf{q}_{hk}$. The allowed wave vectors of the scattered beams are defined by the relations

$$|\mathbf{k}| = |\mathbf{k}^0| \qquad (2.6a)$$

$$\mathbf{k}_{//} = \mathbf{k}_{//}^0 + \mathbf{q} \qquad (2.6b)$$

These conditions can also be derived by the so-called Ewald construction (as in the case of X-ray diffraction) shown in Figure 2.4 for the simple case of normal incidence of the primary beam. Here, a vertical section of the 2D reciprocal lattice is shown. The directions of the scattered beams are defined by the intersections between the sphere of radius k and the vertical lines corresponding to each lattice point. It can thus be realised that the LEED pattern displayed on the fluorescent screen is a kind of projection of the 2D reciprocal lattice. As a consequence, starting from the pattern of the clean surface, the adsorption of atoms or molecules

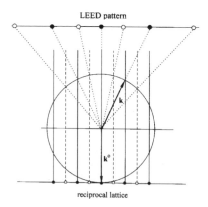

Figure 2.4. Ewald construction for LEED diffraction (normal incidence). Full circles: 2D reciprocal lattice points of the clean surface. Open circles: Additional points due to the presence of a superstructure caused by an adsorbate.

in ordered layers will cause a change in the surface periodicity and new beams will appear in the pattern, as shown in the figure.

Each diffracted beam, corresponding to a given q_{hk}, can be considered as arising from in-phase scattering by the atomic rows with Miller indices (hk). Actually, the conditions for diffraction defined by equations (2.6), correspond to the relation

$$d_{hk}(\sin \theta_{hk} - \sin \theta^\circ) = \frac{2\pi}{k} = \lambda \qquad (2.7)$$

where θ° and θ_{hk} are the angles of the incident and scattered beam, respectively, with respect to the direction normal to the surface; the electron wavelength λ, in Å, is given by

$$\lambda(\text{Å}) = \sqrt{\frac{150}{E(\text{eV})}}$$

where the energy E of the primary electrons is in electronvolts.

Thus, it is possible to derive the distance between atomic rows and the dimensions of the unit mesh from the measurement of the diffraction angles.

The problem of determining the atomic positions within the unit mesh, however, can only be solved by an analysis of the intensities of the scattered beams. Usually, the experimental information is contained in the so-called I/E curves. For each beam, the variation of the intensity as a function of the primary energy E is measured. Then, starting from a possible structural model and using a proper theoretical treatment of the scattering process, intensity curves are calculated and compared with the experimental ones. Using a trial and error procedure, the structure is considered to be solved when a sufficient level of agreement is reached.

If LEED is relatively simple from an experimental point of view, the theoretical treatment of the scattering process is quite complex and is still a serious limitation of this technique. The scattering process for electrons having energies in the range used for LEED is much more complicated to treat in comparison with the scattering of X-rays. In the latter case, the cross-section for the scattering process is so low that the intensity of the scattered waves is a very small fraction of that of the incident one. Thus, the coherent superposition of single scattering events from each atom is sufficient to construct the intensity of a given diffracted beam. On the other hand, in the case of LEED the scattering cross-section is many orders of magnitude higher, so that multiple scattering effects are of importance. This means that the wave acting on each atom is not only that corresponding to the primary electrons, but it also includes the sum of all the waves scattered by the neighbouring atoms (at least those within a distance of the order of the electron mean free path). Thus, a special treatment is needed, called the dynamical theory. Such theoretical treatment was developed in the years around 1970 and later has been applied successfully to the determination of the structure of clean surfaces as well as of adsorbed layers.

The related theory is too complicated to be treated here. The reader can find it in specialised treatises [3,4].

The theoretical treatment introduces some parameters, which are not dependent on the atomic positions, but must be optimised in order to resolve the structure. These are: the inner potential, an effective constant potential supposed to operate in the spaces between the atomic cores, and whose effect is a shift in the electron energy scale; the imaginary part of the potential, simulating an attenuation of the electron waves along their path in the solid; the surface Debye temperature, which introduces the effect of atomic vibrations on the scattered intensities (see Section 4). These parameters are called 'non-structural' as opposed to the 'structural' parameters describing the atomic positions.

Up to a few years ago, due to the complexity of LEED intensities calculations, only relatively simple structures could be solved. At the moment, with the development of new calculation methods and the capability and cost of the computers available, complex structures can be determined by LEED intensity analysis.

The procedure for the determination of the right structural model starts, of course, with the collection of reliable experimental data in the form of intensity vs. energy (or I/E) curves for as many as possible diffraction beams, at several incidence angles of the primary electron beam. From the experimental point of view, the difficulties may derive from stray magnetic fields which can alter the electron trajectories (the Earth's magnetic field must be eliminated in the diffraction chamber, since the electrons at the energies of LEED are very sensitive to magnetic fields) and from inhomogeneities of the fluorescent screen (on which the intensities are measured by a TV camera connected to a computer). Surface preparation is also important, since the smoothest is the prepared surface, the sharper is the LEED diagram and the more reliable are the measured intensities. However, it is found that LEED is not so sensitive to surface defects, like steps or domain boundaries. Recent direct surface images at atomic resolution obtained by means of scanning tunnelling microscopy (STM) (see the following section) have shown that relatively poorly ordered surfaces can give good LEED patterns. This is due mainly to the fact that LEED requires relatively small domains with long-range order (typically a few hundred ångströms wide). The experimental results are then compared with theoretical curves calculated on the basis of different structural models of the surface structure. The right model is that giving the best agreement between experimental and calculated curves. The problem now arises how to judge the degree of agreement. In some cases, the difference between two models is so evident that one of them can be ruled out easily. But not infrequently two different models can both give an acceptable agreement for particular values of the structural and non-structural parameters. In order to make the choice unambiguous and independent of personal opinion, and particularly in order to optimise the values of the parameters and obtain the 'best' structure, several types of 'reliability factors' (R-factors) have been proposed. Some R-factors not only consider the difference in intensity of the curves, but also the difference in the derivatives with respect to the energy, thus increasing the sensitivity to the positions of maxima and minima in the *I/E* curves.

Once a given R-factor is used, its value for a given structural model depends on structural and non-structural parameters. Usually, R-factors are plotted as maps in which one structural parameter (for example, the interplanar distance between first and second atomic planes) is varied together with a non-structural parameter (usually the inner potential). In general, if the structural model is correct, a single, more or less sharp minimum appears in the map, corresponding to the best values of the two parameters. Several maps, where two parameters are varied at a time, allow optimisation of all the parameters of the structure.

An example of determination of a structure with LEED is shown in Figures 2.5 and 2.6. This is the case of a surface alloy formed by deposition of one third of a monolayer of tin on the (111) face of platinum, as studied in reference [5]. The geometry of the LEED pattern indicates the formation of a $(\sqrt{3} \times \sqrt{3})R30°$ superstructure, which can be explained with two different models, as shown in 1.5. In the same figure, the calculated curves for different structural models are also shown. Only one model gives an acceptable agreement with the experimental curves measured for several diffracted beams. Moreover, the correct model, where the Sn atoms are placed among the surface Pt atoms, can be refined by allowing a vertical displacement between the Sn and Pt subplanes. An R-factor map (see Figure 2.6) clearly shows how the best values of both the first interplanar distance and the Sn–Pt sub-planes displacement can be derived.

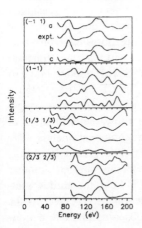

Figure 2.5. Upper part. Two alternative models for the observed $(\sqrt{3} \times \sqrt{3})R30°$ structure formed by deposition of 1/3 monolayer of Sn on the Pt(111) surface: 'Surface alloy' model (Sn atoms incorporated in the topmost Pt layer) (left) and simple overlayer model (right). Lower part. Comparison between the experimental intensity vs. energy curves and those calculated on the basis of various structural models for several LEED beams. Curves labelled a: Surface alloy model. Curves labelled b and c: Simple overlayer model. In this case two possible positions were considered for the non-equivalent threefold sites on the (111) face of an fcc lattice). A visual inspection clearly indicates a better agreement with the surface alloy model [15].

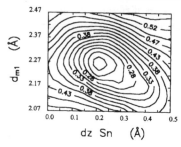

Figure 2.6. R-factor map for the Pt(111)-($\sqrt{3} \times \sqrt{3}$)R30°–Sn surface alloy (see Figure 2.5). d_{m1} is the interplanar distance between the first and second atomic planes and dz Sn is the outward displacement of the Sn atoms with respect to the Pt ones in the first layer. ([15])

At present, the atomic positions can be determined by LEED with an accuracy of about 0.05 Å in the direction perpendicular to the surface, which corresponds to a much better accuracy in the bond lengths.

2.2.2   Other Techniques for Surface Structure Determination

Several experimental techniques can give more or less direct information on the atomic positions at the solid surface. Some of them can be applied to cases that cannot be studied by LEED, since this last technique is not suitable to study surfaces lacking long-range order. An example of a structural technique not requiring long-range order is X-ray photoelectron diffraction (XPD). Photoelectrons emitted under irradiation with monochromatic X-rays typically have kinetic energies of several hundred electronvolts. They are emitted in the form of a wave which is scattered by the neighbouring atoms. Each neighbour becomes the source of a scattered wave and all scattered waves adds to the primary emitted wave. Thus, in a given direction, where the photoemitted electrons are detected, the resultant intensity depends on the spatial distribution of the atoms surrounding the emitting species. Using a single crystal surface, where all the emitting atoms have the same environment, the emitted intensity for a given atomic level will vary as a function of the detection angle. In XPD, the intensity is measured in a wide enough range of azimuthal and polar angles using an electron spectrometer with a small acceptance angle. As for the case of LEED, a comparison of experimental intensity vs. angle curves with curves calculated assuming a given structure allows determination of the correct structural model and optimisation of the structural parameters. The advantage of XPD is that, by choosing the emission from a given atomic level, it is possible to explore the atomic structure surrounding that atomic species. Another advantage is that, working with photoelectrons of 500–1500 eV, the atomic scattering factors are particularly strong in the forward direction with respect to the incoming electrons, so that usually intensity maxima are observed in the directions corresponding to the main atomic alignments. This is often of great help in order to identify the correct structural model. For example, it is immediately clear if an atom is adsorbed at the surface or it is diffused into the bulk. In the first case, an angular variation of the photoelectron intensity is observed only at extreme grazing emergence (typically above 70° from the normal to the surface), whereas in the second case strong intensity variations are measured in the whole angular range.

X-rays have long been considered to be not suitable for surface structure determination, since, due to their deep penetration into the bulk, the contribution of surface atoms to the scattered intensities is too small to be detectable. Recently, however, the use of the strong emission intensity from synchrotron sources, together with a very low background intensity, has allowed the small contribution of surface scattering to be measured, with the advantage of a simpler data analysis, using the usual scattering theory of X-ray crystallography to determine the atomic positions. The main drawback is, of course, the availability of a synchrotron radiation facility, but other difficulties may derive from the very low grazing angles of incidence (of the order of a few tenths of a degree), which require a high perfection in the sample surface preparation.

The most direct structural technique is, however, the scanning tunnelling microscopy (STM). The basis of this technique is the quantum tunnelling of electrons between a sharp tip and the surface, when the tip is kept at a few ångströms distance. The tunnelling is possible since the electron wave function does not drop to zero at the surface (see Section 3.1) but it decays exponentially within a few ångströms. Thus, if two conductors are approached to such a distance, the overlap of the two electron densities will allow an electric current to be measured if a small potential difference is applied. In STM the probe tip is mounted on a piezoelectric holder which, under the application of a suitable potential, allows it to be moved along the surface with a very high precision. Employing fast-response electronic feedback circuits, which keep the tunnelling current constant, the tip can be held at this close distance while it is scanned along the surface. Thus, any local variation of the charge density causes a variation in the circuit controlling the tip distance. If these variations are plotted on a screen, a real image of the surface is obtained, with the possibility of reaching atomic resolution.

STM has allowed several complex surface structures to be solved. A typical example is the (7×7) reconstruction which is observed on the clean Si(111) face after annealing at high temperatures. This superstructure could not be resolved by LEED because of the high number of atoms per unit mesh, so that several models were proposed to explain the (7×7) periodicity, until the application of STM showed which was the correct one. STM is very useful not only to solve the case of complex structures but also for the study of surface defects and of their movement as a function of the temperature, as well as for the study of the nucleation and growth of islands of a new solid phase. An example is the formation of CuCl islands on a Cu(100) face covered with adsorbed Cl atoms (Figure 2.7).

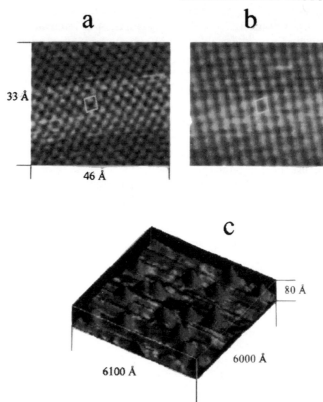

Figure 2.7. STM images of a Cu(100) face exposed to chlorine: (a) clean surface; (b) surface covered by a monolayer of chlorine atoms forming a c(2 × 2) structure. Image size: 33 Å × 46 Å. In both cases the square drawn on the figure indicates the c(2 × 2) unit cell; (c) formation of CuCl islands (Image size: 6100 Å × 6000 Å × 80 Å) (Courtesy of K. Elsov *et al.*).

When the copper surface is exposed to gaseous chlorine at pressures of the order of 10^{-7} torr, it is first covered by a monolayer of chlorine atoms packed in a c(2 × 2) structure. Upon further exposure, islands of CuCl are formed. The composition of the new phase was determined with XPS (see Section 5) while its structure and epitaxial relations with the copper surface were determined with XPD.

At present, STM is becoming a routine microscopy for the study of surfaces at the atomic scale. However, it must be taken into account that, if STM is capable of giving a direct image of the surface, it cannot detect atomic positions below the first atomic plane. Moreover, it is in general not able to distinguish between two different atomic species, unless they differ significantly in size or electron density. Thus, STM is not suitable to determine interatomic distances and surface composition. Another kind of problem is the interaction of the tip atoms with adsorbates, that may be an obstacle for the study of adsorbed molecular species.

2.3 MAIN RESULTS OF SURFACE STRUCTURE DETERMINATIONS

The number of surface structures already studied has rapidly increased since the development of LEED and, successively, of other structural techniques, so that only general results will be reported here. For a recent bibliography of structures investigated, see references [6] and [7].

2.3.1 Clean Surfaces

The study of the structure of the clean surface of a solid is obviously preliminary to any investigation on surface phenomena. At present, due to the availability of high purity single crystals, almost all solid surfaces can be studied by LEED or other structural techniques. It must be taken into account, however, that most of the techniques available for surface structure determinations require a high vacuum environment. In any case, high or ultra-high vacuum conditions are necessary to prepare and maintain a clean surface. This requirement may cause problems in the case of solids with a relatively high vapour pressure, or whose stoichiometry can be changed under vacuum. A typical case is that of some oxides which, during the initial cleaning procedure under vacuum (often requiring annealing at high temperatures to recover the long-range ordering damaged by a polishing treatment or an ion bombardment), can lose oxygen, thus becoming non-stoichiometric at the surface.

One of the most significant results of surface structure investigation, which has become apparent since the beginning of LEED studies, was the demonstration that in some cases the structure of a clean solid surface may differ significantly from that of the parallel planes in the bulk (i.e., the ideal structure). A change in the

Table 2.2. Clean surfaces showing reconstruction

Au, Pt, Ir	$(100)\text{-}(5 \times 1)$; $(110) \times (1 \times 2)$
Pd	$(110)\text{-}(1 \times 2)$
W	$(100) \times c(2 \times 2)$ (below room T)
Bi	$(11\text{-}20) \times (2 \times 10)$
Sb	$(11\text{-}20) \times (6 \times 3)$
Si	$(111)\text{-}(2 \times 1)$; $(111) \times (7 \times 7)$
	$(100)\text{-}(2 \times 1)$; $(100)\text{-}c(4 \times 2)$; $(110) \times (5 \times 2)$
Ge	$(111)\text{-}(2 \times 1)$; $(111) \times (2 \times 8)$
	$(100)\text{-}(2 \times 1)$; $(100)\text{-}c(4 \times 2)$; $(110) \times (2 \times 1)$
GaAs, GaP, GaSb, InSb	$(111) \times (2 \times 2)$
C (diamond)	$(111) \times (2 \times 2)$

surface unit mesh (reconstruction) was observed for several solids. Representative examples of surface reconstruction are reported in Table 2.2.

Most metal surfaces do not show reconstruction, with a few exceptions. In the case of the (100) faces of Au, Pt, and Ir the reconstruction is mainly due to a few percent contraction of the interatomic spacing in the first plane, leading to a more stable hexagonal 2D packing that forms a 'coincidence' superstructure with the second unmodified plane, showing an apparent (5×1) pattern in the LEED diagram (actually, for Au and Pt the coincidence mesh is more complicated). The (110) faces of these metals show a (1×2) reconstruction that has been interpreted with a 'missing row' model, where every other top layer row parallel to the $(1\bar{1}0)$ direction is missing.

The W(100) face shows a reconstruction below room temperature, consisting of small lateral displacements of the top layer atoms with formation of zig-zag chains.

For all metals, however, a more or less pronounced contraction of the first interplanar spacing has been found. This variation is stronger for faces with a less compact structure. Thus, contractions of only a few percent have been reported for the (100) faces of fcc metals, while variations of the order of 10–15% were found for the more open (110) faces. The most closely packed (111) faces of fcc and the (110) of bcc metals are almost unchanged. The interplanar spacing between the second and third atomic planes appears to be only slightly changed, a small expansion of a few percent being found for the cases of strong top layer contraction.

For semiconductors, on the other hand, the nature of the bonds is covalent and the loss of neighbours at the surface is more difficult to compensate for without significant displacements of atomic positions. This is the reason for the reconstructions observed for most semiconductor surfaces. In this case the perturbation of the structure may propagate several atomic layers into the bulk. The (2×1) reconstruction of the Si(100) face is due to a pairing of top layer atoms with lateral displacements of 0.4–0.8 Å, the two atoms of the pairs being not coplanar. The case of the Si(111)–(7×7) superstructure has been solved only by using STM, and it is due to the presence of periodic vacancies which favour a better rearrangement of the dangling bonds on the remaining Si atoms in the top layer. By means of other techniques, it was found that slight variations of atomic positions extend for several atomic planes inside the solid.

Several inorganic compounds have been studied, such as oxides, sulphides, halides, etc. In general, only their cleavage faces have been investigated. These do not show any reconstruction or significant variations of the first interplanar distance. Only in the case of a few oxides, a variation in the surface periodicity has been observed and it has been attributed to a change in surface stoichiometry.

2.3.2 Surfaces with Adsorbates

When atoms or molecules are adsorbed on a previously clean surface, the LEED diagram immediately shows whether the adsorption layer has long-range order or is partially or completely disordered. One of the first and most striking results of LEED was the direct evidence of formation of ordered phases with different structures when the surface coverage of a given adsorbate is varied. The capability of LEED to directly show on a fluorescent screen the formation of ordered adsorption structures (usually corresponding to well-defined adsorbate coverages) is the reason why it is still the most widely used technique for surface structure investigation. When ordered structures are formed, the surface unit mesh can be determined, at least when the relation between the periodicity of the layer and that of the support is fairly simple. The knowledge of the unit mesh is fundamental in order to define the exact concentration of the adsorbed species. Sometimes, the identification of the unit mesh is not so straightforward, due to the existence of several equivalent orientations of the layer with respect to the substrate. A simple example of this ambiguity is that of a (2×1) superstructure on a (111) face of a cubic crystal: in this case, equivalent ordered domains can exist which are rotated by 120°. If the domain size is much smaller than the electron beam diameter (which is usually the case), the LEED diagram is the superposition of three (2×1) patterns rotated 120°, that is, the same as the pattern observed for a (2×2) superstructure. The interpretation of the LEED pattern can be even more complex with larger unit meshes, particularly when the unit cells are rotated with respect to the substrate lattice. In this case, multiple diffraction effects may give rise to additional diffracted beams, thus further complicating the interpretation of the pattern geometry.

When the surface unit mesh is not too large, the dynamical theory can be applied to calculate the LEED intensities, so that the bond distances and angles can be derived. A great number of structures formed by atomic adsorbates has been determined by LEED and, more recently, with the growing contribution of the other

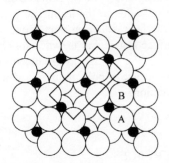

Figure 2.8. Model for the $(\sqrt{2} \times 2\sqrt{2})$ structure formed by oxygen adsorption on the Cu(100) face, as determined by LEED. Black circles represent oxygen atoms (almost coplanar with the copper top layer). Oxygen atoms are found about 0.1 Å above and copper atoms marked 'A' 0.5 Å below the plane of 'B' atoms. Moreover, the A atom rows are laterally displaced by about 0.1 Å from the B rows. The average vertical position of A and B atoms with respect to the second atomic plane is 0.25 Å higher than the bulk interplanar distance, corresponding to a relaxation of 14% ([19]).

structural techniques. A review of the structures resolved can be found in references [6] and [7].

In most cases, it has been found that atoms are adsorbed on metal surfaces in hollow sites with the highest coordination, with minor structural changes of the top atomic layers, that is, a variation of the first interplanar distance or small lateral displacements around the adsorption site. In several cases, however, a surface reconstruction induced by chemisorption has been observed. For example, oxygen adsorption on Cu(100) produces a $(\sqrt{2} \times 2\sqrt{2})$ structure where one in every four copper rows parallel to the [010] direction is missing. This way, deeper adsorption sites are available for the oxygen atoms (see Figure 2.8). A full dynamical LEED study (see reference [19]) allowed small lateral shifts of the copper atoms to be found and the vertical positions of the oxygen atoms to be determined.

Relatively few structures of molecular adsorbates have been resolved as yet. This has been mainly caused by the difficulties encountered with LEED, since the adsorbed molecules can desorb or decompose under the electron beam. This requires the use of special LEED equipments with very fast measurements of the intensities or with very low electron currents and special detectors to amplify the scattered beams. In this field, the contribution of the other techniques, particularly those using synchrotron radiation, is rapidly increasing (see references [1], [6] and [7]).

3   ELECTRONIC PROPERTIES OF SURFACES

Electronic states, charge density and bonding states at surfaces are to some extent different from those of the bulk due to the asymmetric environment of the atoms close to a vacuum–solid interface. Electron wave functions are modified by the presence of the surface: surface states, that is, wave function solutions localised at the surface, become possible. The change in the electron distribution around the surface atoms produces a surface dipole moment which determines important properties like the work function.

3.1   ELECTRONS AT SURFACES

Electronic properties of surfaces can be described in the framework of the band structure model of solids. Electrons in solids occupy allowed energy bands separated by forbidden energy regions which correspond to values of the wave vector **k** of electrons for which no propagating-wave solutions exist. These energy bands are obtained from solutions of the Schrödinger equation for an electron moving in the periodic potential due to an infinite lattice of ion cores and to an average potential produced by the charge distribution of all other valence electrons. Wave functions in a periodic potential can be written on the basis of the Bloch theorem as

$$\psi_k(\mathbf{r}) = u_k(\mathbf{r})\exp(i\mathbf{k} \cdot \mathbf{r}) \tag{2.8}$$

where $u_k(\mathbf{r})$ is a function with the same periodicity as the lattice.

When the infinite solid is terminated by the introduction of a surface the wave functions inside the crystal are modified. Additional solutions of the Schrödinger equation can thus be obtained introducing the appropriate boundary conditions at the surface. New electronic states become possible in the energy gap of the infinite crystal: they have complex wave vectors (which are not allowed in an infinite solid because their wave function amplitude would not have a limited value over all space) and their wave functions decrease exponentially from the surface to the bulk. These surface states are delocalised over the whole surface and are characterised by well-defined wave vectors $\mathbf{k}_{//}$ parallel to the surface. The surface states deriving from the simple truncation of the three-dimensional periodic structure of the solid are called Shockley states. Surface electronic states which can be introduced by a perturbation of the potential at the surface are called Tamm states. Tamm electronic states are typical of transition metal and semiconductor surfaces. In the case of semiconductors, due to the directional character of the bonds between atoms, surface electronic states can be seen as dangling bond orbitals (localised on the surface atoms) which are produced by a cleavage of the crystal.

Surface states can also be introduced by changes in the potential due to relaxation, reconstruction and structural defects of the surface. Moreover, surface states can be associated with chemical bonds between adsorbed species and surface atoms. Surface states can determine the chemical reactivity of the surface since they can act as donor or acceptor levels.

The presence of surface electronic states has an effect on the charge density at the surface. If such states are present below the Fermi energy, they can be occupied by electrons. In this process, an excess negative charge is accumulated at the surface. Therefore, due to presence of surface states, an electric double layer is present at the surface of a solid. This electrostatic dipole layer at the surface can be evaluated in the case of s–p metals, like alkali metals or aluminium, by means of the so-called 'jellium model' of the surface. In this model the discrete ion cores are replaced by a uniform, positive charge distribution with density equal to the spatial average of the core ions charge. The electron density variation perpendicular to the surface derived from the jellium model shows that there is no sharp edge to the electron distribution but the electrons tend to spill out into the vacuum region.

Calculations of the surface electronic structure of metals and pure semi-conductors, performed using various approaches, show that bulk properties are recovered beyond a few atomic layers inside the crystal. However, in the case of doped semiconductors the presence of surface states leads to the formation of a space charge region that can extend up to several thousand ångströms inside the solid modifying the band structure of the crystal at the surface (see Section 3.5).

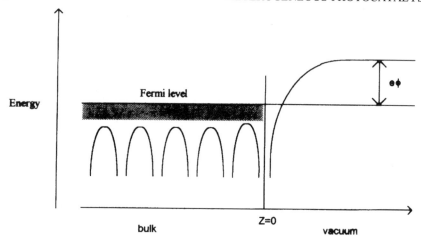

Figure 2.9.   Energy of an electron close to the surface of a solid.

3.2   WORK FUNCTION

The work function $e\phi$ is defined as the energy required to remove an electron at the Fermi level to the free space outside the solid. For a metal, the work function corresponds to the difference in energy between an electron at rest in the vacuum just outside the solid and an electron in the highest occupied level in the solid at 0 K (Figure 2.9). The work function is an electronic property which is relevant in all processes involving the emission of electrons from surfaces, like thermionic emission, electron emission induced by photons (photoemission) and field emission.

There are bulk and surface contributions to the work function. The bulk contribution comes from the attraction of the electrons from the ion cores of the lattice. If the charge density of the surface layer were the same as that of the bulk and no perturbation of the charge density were induced by the escaping electron, this would be the only contribution to the work function. This is not the case since the actual charge distribution in the surface plane differs from that of the planes in the interior and it is perturbed by the outcoming electron. Thus, there are surface specific contributions to the work function.

In the presence of a high concentration of free electrons, as in the case of a metal, one surface term comes from the work necessary for an escaping electron to cross the dipole layer present at the surface. The other surface contribution is related to the image charge potential. The attractive potential that an electron experiences when it is beyond a few lattice spacings from the surface can be described as the electrostatic potential between the negative charge of the electron at distance z from the surface outside the solid and the image positive charge at distance $-z$ inside the solid. The force due to the image charge is negligible beyond a few thousand ångströms away from the surface.

Due to the surface contributions, the work function of a solid depends on the crystallographic orientation of the surface plane and on the presence of adsorbates. The dipole moment at the surface depends upon the density of the atoms at the surface, which varies with the crystallographic orientation. For this reason, the most densely packed surfaces of solids tend to have the highest work function. Also the presence of steps on the surface has an effect on the work function, a stepped surface having a lower work function than a flat one. The reason is that electrons spill out at the surface but they tend to smooth the sharp step along the surface. The result is an electric dipole oriented in the opposite direction to the dipole of the surface layer, thus reducing the net dipole moment, and hence the work function, relative to the flat surface value.

When atoms or molecules are adsorbed on the surface, the work function changes because of the modification of the electron density at the surface. The work function increases if there is a negative charge transfer from the substrate to the adsorbate. This is the case of adsorption of electronegative species (for instance, oxygen and halogens) on metal surfaces. The work function decreases if a charge transfer occurs from the adsorbate to the substrate, as in the case of adsorption of alkali metals on transition metal surfaces. For molecular adsorbates, the change depends on the dipole moment of the adsorbate as well as on the charge transfer upon adsorption. For instance, carbon monoxide causes an increase of the work function, while hydrocarbons usually cause a decrease when adsorbed on transition metals.

The work function can be directly measured in photoemission experiments (see Section 3.3) or its changes can be monitored by measuring variations of the contact potential with respect to a reference electrode (vibrating capacitor method).

3.3 PHOTOEMISSION AS A PROBE OF THE SURFACE ELECTRONIC STRUCTURE

Photoemission, in particular angle-resolved photoemission, is one of the most powerful techniques for studying the electronic properties of surfaces. In a photoemission experiment electrons emitted from the surface region of a solid upon irradiation with a monochromatic electromagnetic radiation are analysed in energy and in angle. The kinetic energy of an electron emitted from a given electronic level by an incident photon of energy $h\nu$ is related to the binding energy E_b (or ionisation potential) of that level by the following equation:

$$E_{kin} = h\nu - E_b \tag{2.9}$$

Within a HF–SCF treatment and assuming that the removal of an electron does not introduce any change in the remaining electrons, the binding energy of an electron in a given orbital is equal to the negative of the orbital energy (Koopmans' approximation). This approximation is expected to better hold for valence levels since in this case relaxation effects are negligible. For core levels, the relaxation in the final state may be relevant (it is about 14 eV for C 1s).

In the case of photoemission from solids, referring the binding energy to the Fermi level we obtain

$$E_{\text{kin}} = h\nu - E_{\text{b}}^{\text{f}} - e\phi \tag{2.10}$$

where E_{b}^{f} is the binding energy referred to the Fermi level and $e\phi$ is the work function of the sample. Of course, due to the contact potential between the sample and the electron analyser, the effective kinetic energy measured by the instrument (see Figure 2.10) will be

$$E_{\text{kin}}' = h\nu - E_{\text{b}}^{\text{f}} - e\phi_{\text{a}} \tag{2.11}$$

where $e\phi_{\text{a}}$ is the work function of the analyser. When working on solids, the advantage of referring the binding energies to the Fermi level is that it is not necessary to know the work function of each sample, since E_{b} can be obtained from the measured kinetic energy if the analyser work function is known from a suitable calibration of the energy scale.

Depending upon the energy of the photons used, we can divide photoemission techniques into two groups: ultraviolet photoelectron spectroscopy (UPS) and X-ray photoelectron spectroscopy (XPS).

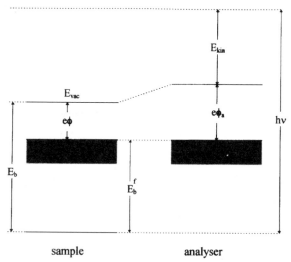

sample   analyser

Figure 2.10. Schematic energy level diagram for photoemission from solids. A photon of energy $h\nu$ is adsorbed by an electron in the level E_{b} below the Fermi level E_{f}. The kinetic energy of the emitted electron is given by equation (2.10). E_{b} and E_{b}^{f} are the binding energies referred to the vacuum level and to the Fermi level, respectively. ϕ and ϕ_{a} are the work functions of the sample and of the analyser, respectively.

UPS uses photons in the vacuum ultraviolet region of the electromagnetic spectrum (i.e., a range from 10 to 100 eV), whereas in XPS photons in the X-ray range (1–2 keV) are employed. In both cases, the escaping depth of the emitted photoelectrons is of the order of 5–30 Å, which is the reason why these techniques are so sensitive to the properties of the first atomic planes and are among the most widely used for the study of solid surfaces. In this section, we will deal with UPS as a technique for studying the valence band of solids. XPS, which involves the ionisation of core levels, and its application to the determination of surface composition will be treated in Section 5.

By means of XPS it would also be possible to study the valence band. However, at the typical X-ray photon energies used in XPS, the cross-section for photoemission from valence band levels is very low. Moreover, emitted electrons have relatively high kinetic energy and hence they are less sensitive to the electronic surface states. UPS is a more suitable technique to study the valence band of surfaces because of the higher surface sensitivity (due to the relatively low kinetic energy of the photoemitted electrons) and the larger cross-section for the photoemission process.

Discharge lamps producing resonance radiation lines from He (He I at 21.2 eV and He II at 40.8 eV) are the most common used photon sources in laboratory-based UPS experiments. Synchrotron radiation in the VUV range can be used in photoemission with the great advantages of being tunable and polarised. Hemispherical analysers are used in UPS to separate electrons having different kinetic energy.

A typical UPS spectrum of a metal surface is shown in Figure 2.11. The right end of the spectrum (the zero of the binding energy scale) corresponds to electrons

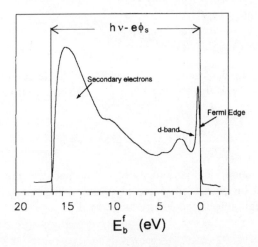

Figure 2.11. UPS spectrum of the Pd(100) surface excited by He I radiation (21.21 eV).

emitted with the highest kinetic energy, that is, electrons emitted from the Fermi level, with a kinetic energy $E_{kin} = h\nu - e\phi_S$. Emission down to some electronvolts below the Fermi energy displays the density of states of the metal valence band. The continuous background, increasing at lower energies, is due to secondary electrons produced by energy losses of the photoelectrons during their path to the surface. The cut-off of emission on the left end of the spectrum corresponds to electrons having zero kinetic energy. Thus, from the width ΔE of the UPS spectrum it is possible to determine the work function of the surface:

$$e\phi_S = h\nu - \Delta E \tag{2.12}$$

In the energy distribution curve of the photoemitted electrons some features are present which can be attributed to electrons emitted from bulk electronic states. In the case of adsorbed atoms or molecules emission from molecular orbitals of the adsorbate can be observed in the UPS spectra. The comparison of photoemission spectra of molecules in the gas phase and in the adsorbed phase can give information about the nature of the surface chemical bond. By means of UPS spectra it is possible to determine in a straightforward way if a molecule decomposes upon adsorption at a given temperature. A typical example is the case of the UPS studies of CO adsorbed on transition metal surfaces.

More details about electronic states can be obtained from angle-resolved UPS. By means of this technique it is possible to map out the surface band structure of a solid, that is to obtain the wave vector dependence of the band energy. On the basis of equation (2.10) and of Koopman's approximation, the energy of the initial state from which an electron is emitted can be determined. The electronic states are characterised by wave vectors parallel to the surface $\mathbf{k}_{//}$ which can be determined from momentum conservation. The kinetic energy of an emitted electron is related to the wave vector components parallel and perpendicular to the surface by the following equation:

$$E_{kin} = \frac{\hbar^2}{2m}(|\mathbf{k}_\perp^0|^2 + |\mathbf{k}_{//}^0|^2) \tag{2.13}$$

where $\mathbf{k}_\perp^0$ and $\mathbf{k}_{//}^0$ are determined by the position of the analyser with respect to the surface of the sample. If θ is the angle between the surface normal and the electron analyser

$$|\mathbf{k}_{//}^0| = \left(\frac{2mE_{kin}}{\hbar^2}\right)^{1/2} \sin\theta \tag{2.14}$$

Due to the two-dimensional periodicity of the surface, the momentum parallel to the surface must be conserved in the photoemission process. Since the photon momentum is negligible in the energy range used in UPS, the $\mathbf{k}_{//}$ wave vector of the initial state is given by

$$\mathbf{k}_{//} = \mathbf{k}_{//}^0 - \mathbf{q} - \mathbf{g}_{//} \tag{2.15}$$

where $\mathbf{q}$ is a lattice of the two-dimensional reciprocal lattice and $\mathbf{g}_{//}$ is the component parallel to the surface of a vector of the bulk reciprocal lattice (see Section 2).

The perpendicular component of the photoelectron wave vector bears no particular relation to $\mathbf{k}_{\perp}$ of the initial state.

If the energy distribution curves of emitted electrons are collected at a fixed photon energy for a series of emission directions, peaks in these curves correspond to initial states indexed by $\mathbf{k}_{//}$. The variation of the binding energy of an electronic state as a function of the detection direction represents the dispersion relation $E(\mathbf{k}_{//})$ of that electronic state. Dispersion curves can be obtained also collecting spectra at a certain emission direction for different photon energies. The set of dispersion relations obtained from angle-resolved UPS (ARUPS) can be compared with band structure calculations. If polarised radiations is used in a photoemission experiment, selection rules based on symmetry considerations allow one to obtain useful information about the initial state. An example of a study of the dispersion relation of the surface states using ARUPS is that of reference [8] (see also references [9] and [10]).

3.4   ELECTRONIC PROPERTIES OF METAL SURFACES

In the case of s—p metals, the atomic cores scatter relatively weakly the s and p valence electrons. Therefore, the energy bands formed from s and p atomic orbitals are very similar to the free electron bands. For simple metals, surface electronic properties can be derived on the basis of the jellium model (see Section 3.1). Indeed, more rigorous band structure calculations for s—p metal surfaces confirmed the results derived from simple models. Surface electronic states of simple metals are Shockley states.

The electronic properties of transition metals are characterised by the conduction bands formed by the overlapping of the more localised d orbitals. In the case of transition metal surfaces, Tamm surface electronic states are present. One of the interesting properties of transition metals is the d band narrowing at the surface. From tight binding calculations it turns out that the width of the local density of states at a given site is proportional to the coordination number of that site. This means that the local density of states is narrower at the surface than in the bulk since surface atoms have fewer neighbours than bulk ones. Due to the d band narrowing at the surface, charge transfer must occur between surface atoms and atoms in the bulk in order to maintain a common Fermi level. If the d band is less than half-filled, charge will flow from surface atoms to the bulk. The direction of charge transfer is reversed in the case of more than half-filled. This variation of charge in the valence orbitals of the atoms at the surface produces a different electrostatic potential at the surface with respect to the bulk. The effect of this charge transfer can be probed, experimentally by means of XPS: it appears as a shift in the core levels of surface with respect to bulk atoms.

3.5 ELECTRONIC PROPERTIES OF SEMICONDUCTOR SURFACES

The electronic structure and properties of semiconductor surfaces can be qualitatively described in terms of chemical bond concepts. In the diamond- type structure and in the zincblende type structure (the most common structures of semiconductors) each atom is bonded to its four tetrahedrally coordinated neighbours by sp^3 hybrid orbitals. When the three-dimensional structure of the semiconductor is cut by a surface, dangling bonds are formed. The surface state wave functions, which extend over the surface and have a well-defined wave vector parallel to the surface plane, can be seen as linear combinations of these dangling bond orbitals. When the (111) surface of a diamond-type structure is formed, one bond is cut for every surface atom. Therefore, there is one band of surface states. In the case of the (100) surface, two bonds are cut for every surface atom and, since there is one atom per unit cell, two bands of electronic state are expected. These considerations hold for ideal bulk truncated surfaces. However, this is not the case for the low index surfaces of silicon and germanium which are all reconstructed in order to eliminate or reduce the number of dangling bonds. For these surfaces other electronic states, related to reconstruction and relaxation, are present besides dangling bond states.

At the surface of pure semiconductors, surface states in the band gap remove charge density from the top of the valence band. Valence band charge deficit is compensated by charge in the surface states. This variation of the charge density is on an atomic length scale.

The situation is different in the case of surfaces of semiconductors doped with impurities. For an n-type semiconductor, electrons from the donor levels close to the surface region are transferred to the surface states at the mid-gap until an equilibrium condition is achieved. In this process a space charge region forms which may extend several hundred or thousand ångströms into the bulk because the concentration of impurities per unit area is much smaller than the density of surface states (the latter is typically of the order of 10^{14} cm^{-2}; thus, for a semiconductor with a doping of 10^{19} impurity atom cm^{-3}, a slab of about 10^3 Å would be involved in the space charge region). The electrostatic potential produced by the space charge region causes a modification in the band structure near the surface, which is known as 'band bending'. For n-type semiconductors, the bands rise up at the surface (see Figure 2.12). In a p-type semiconductor the bands bend downward, since electrons are transferred from the surface states to the acceptor levels.

For semiconductor surfaces it is useful to define two other parameters in addition to the work function. One is the band bending V_S, that is, the energy shift of the bands at the surface with respect to the bulk. The other is the electron affinity χ, which is the energy needed to remove an electron from the bottom edge of the conduction band at the surface.

An effect of surface electronic states and band bending is that the work function of a doped semiconductor is nearly independent from the dopant concentration.

n-type semiconductor

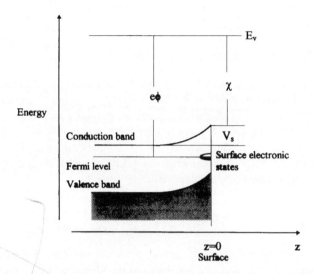

Figure 2.12. Schematic representation of the band bending for an n-type semiconductor surface.

Doping a semiconductor with a donor impurity raises the Fermi level and the work function should decrease with increasing the dopant concentration. In fact, this effect is cancelled by the increase in the energy spent by an escaping electron to cross the space charge region and the work function is nearly independent from the dopant concentration. This phenomenon is known as pinning of the Fermi level for doped semiconductors.

For more detailed information on the surface electronic properties of solids and on the application of UPS, see the general references [10] and [11].

4 SURFACE DYNAMICS

So far, surface properties have been discussed in terms of rigid lattices of atoms. In order to take into account the thermal vibrations of atoms, we should consider that vibration properties characteristic of the bulk are modified at surfaces, since the forces acting on a surface atom are very different from those acting on a bulk one. In particular, we can expect that the oscillations in a direction normal to the surface plane are significantly larger for an atom in the top atomic plane at the surface. We can also expect that the energy required to remove an atom from the top atomic plane, thus creating a surface vacancy, is significantly lower than that needed to create a vacancy in the bulk, which is required to allow the movement of bulk atoms. Moreover, once an atom of the solid is put onto the surface, it can migrate much more easily than an interstitial atom or a vacancy in the bulk lattice.

4.1 ATOMIC VIBRATIONS AT SURFACES

In X-ray crystallography, it is well known that the intensity of diffracted beams decreases as the crystal temperature is raised. From the temperature dependence of the diffracted intensity, it is possible to derive the mean square amplitude of atomic vibrations. Similarly, using LEED, mean square vibration amplitudes of atoms in the surface region of a solid can be determined. Neglecting multiple scattering effects, the intensity of the (0, 0) beam (corresponding to specular reflection of the primary beam) as a function of the temperature is given by

$$I(T) = I_0 \exp\left[-\frac{12h^2}{mk} \left(\frac{\cos \alpha}{\lambda}\right)^2 \frac{T}{\Theta_D^2} \right] \qquad (2.16)$$

where I_0 is the intensity for the rigid lattice, m is the mass of the atoms, k is Boltzman's constant, α is the angle of incidence of the electrons with wavelength λ (see Section 2). Θ_D is an effective Debye temperature of the surface region. Since LEED probes not only the topmost layer, the Debye temperature derived from LEED represents an 'effective' value which tends to the bulk value on increasing the electron energy, i.e., on increasing the penetration depth of the electrons in the solid. In the harmonic approximation, the mean square amplitude (m.s.a.) of atomic vibrations $\langle n^2 \rangle$ is related to the Debye temperature by the equation

$$\langle u^2 \rangle = \frac{3h^2 T}{4mk\pi^2 \Theta_D^2} \qquad (2.17)$$

From the temperature dependence of the (0, 0) beam, only the component of the mean square displacement normal to the surface $\langle u_\perp^2 \rangle$ can be determined. The parallel component $\langle u_{//}^2 \rangle$ can be obtained from the temperature dependence of non-specular beams. The information about the vibration amplitudes derived from LEED can be considered only semiquantitative due to the simplified model used to treat the electron scattering from a vibrating lattice. Nonetheless, from diffraction measurements it turns out that $\langle u^2 \rangle$ is much higher at the surface that in the bulk. In Table 2.3 representative values of Θ_D determined by LEED for several surfaces are shown. Intuitively, on the basis of the classical harmonic oscillator model, since the m.s.a. of vibration is inversely proportional to the force constant, the increase in the m.s.a. of the first atomic layer in the direction normal to the surface is due to the absence of nearest neighbours on the vacuum side. Considering the increased mean square displacements and the corresponding decrease in the vibration frequencies,

Table 2.3. Representative values of surface Debye temperatures determined by LEED and comparison with bulk values (from reference [4])

Surface	Surface Θ_D (K)	Bulk Θ_D (K)
Al(100)	189	370
Ni(110)	220	390
Pd(100)	140	274
Pd(111)	140	274
Ag(111)	155	225
Cu(111)	244	322
Cu(100)	210	322
Pt(100)	118	234
Pt(111)	111	234
W(110)	200	280
W(100)	150	280

anharmonicity effects are expected to be important for thermal vibrations at surfaces.

Speaking in terms of collective atomic oscillations (phonons), in a semi- infinite crystal, there are collective oscillation modes whose vibrational amplitudes decrease exponentially from the surface to the bulk (surface phonons). The vibrational properties of a surface are described by means of phonon bands which give the dependence of the energy of a phonon on its wave vector parallel to the surface.

Vibrational properties are strongly affected by reconstruction, relaxation and adsorption phenomena since these processes involve a variation of the interatomic potentials. Vibrational amplitudes and dispersion relations of surface phonons can be calculated by means of lattice dynamics methods on the basis of the atomic structure and of the interatomic potentials for the surface layers. If the atomic structure is known from other techniques, the fitting of experimental vibrational data with calculations allows one to get information about interatomic forces at the surface of a solid.

One of the main experimental techniques to study atomic vibrations at the surface is high resolution electron energy-loss spectroscopy (HREELS). Electrons scattered from the surface of a solid can lose energy by exciting surface vibrations. Since typical energies of phonons are of the order of few tens of millielectronvolts (1 meV = 8 cm^{-1}), high resolution is needed to detect electrons which have lost energy to excite surface phonons. This extremely high resolution is achieved using a strictly monochromatic electron beam (with energy of the order of a few electronvolts and a band width of 1–2 meV) produced by means of an electron energy analyser. The electrons scattered by the surface are energy analysed by a second dispersive analyser. For more information on HREELS, see reference [9].

Another technique to study surface vibrations is the inelastic scattering of atomic He beams. This technique allows one to have much higher energy resolution and surface sensitivity, but it requires a more complex and expensive experimental equipment.

4.2   SURFACE DIFFUSION

The migration of atoms along the surface may play an important role in several surface phenomena involving the transport of atoms, like the nucleation and growth of a new phase. In surface reactions, the movement of surface atoms and of adsorbed molecules may be one of the most important steps.

In the bulk solid, the movement of an atom from a lattice site to a nearest neighbour one is only possible if the latter is vacant (the alternative mechanism, that is, the direct exchange of two neighbouring atoms, would require too much energy to play a significant role at temperatures lower than the melting point). Thus the diffusion process in the bulk is limited by the availability of lattice vacancies, whose equilibrium concentration increases exponentially on increasing the temperature. For example, the diffusion coefficient for a pure metal in its own lattice (self-diffusion) is given by

$$D = D_0 \exp[-(E_{f,v} + E_{m,v})/RT] \tag{2.18}$$

where $E_{f,v}$ is the energy of formation of lattice vacancies, and $E_{m,v}$ is the activation energy for their migration from a site to another. In a simple treatment of the diffusion process based on the random walk model, D_0 is given by $va^2/2m$, where v is the vibration frequency of the atoms, a the distance between two nearest neighbours and m is the number of directions along which the atom can jump with equal probability. Using a simple broken bond model, the energy required to form a vacancy in an fcc metal would be $6E_b$, where E_b is the energy of a single bond. For example, in the case of copper it would be about 80 kcal/mol (actually it is much lower, because a significant part of the energy is recovered by the relaxation of the lattice around the vacancy). For self-diffusion on the surface, the overall process may involve a much lower energy. What is required is that an atom of the topmost layer is displaced onto the surface plane in a hollow site, thus being free to move from a site to another with a very small energy barrier corresponding to the saddle point between two adjacent sites.

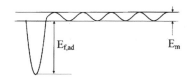

Figure 2.13.   Schematic energy diagram for an atom jumping from a surface lattice site and migrating as an adatom on top of the surface atoms.

This adsorbed atom is called an 'adatom'. Now, the energy required to form an adatom depends on the surface structure. On the (100) face of an fcc metal, the energy required to form an adatom sitting in a fourfold hollow site would be $3E_b$, thus less than the energy of formation of bulk vacancies, while the energy barrier between two adjacent sites passing through a bridge site should be lower than E_b (for copper $E_b = 13$ kcal/mol, to be compared with the observed $E_{m,v} = 21$ kcal/mol for self- diffusion in bulk copper). However, on a real surface there is a great number of steps. Thus, for the (100) face the energy of formation of the same adatom starting from an atom sitting at a step would be only $3E_b$. Moreover, adatoms can even more easily be formed and move along the steps. Therefore, without considering the pre-exponential terms, at a given temperature we can expect a much higher atomic mobility at the surface in comparison with the bulk.

For atoms or molecules adsorbed on a solid surface, the mobility along the surface is limited only by the energy barrier found at the saddle points between two adjacent equivalent adsorption sites. This energy barrier depends on the surface structure and on the type of interaction between the adsorbate and the surface atoms, but it is usually a small fraction of the adsorption energy. Therefore, many adsorbates can have a relatively high mobility even at room temperature.

To have an idea of the mobility of an adatom, we can use a simple treatment giving for its jumping frequency

$$f = zv \exp(-E_m/RT) \tag{2.19}$$

where z is the number of equivalent neighbouring sites and E_m is the energy barrier for migration. Assuming $v = 10^{13}$ and $z = 6$ (as in the case of a (111) face of an fcc structure) at 300 K f would be 3×10^6 jumps/s for $E_m = 10$ kcal/mol. Since the mean square distance travelled by a diffusing molecule in the time t is

$$\langle r^2 \rangle = fa^2 t \tag{2.20}$$

the root mean square distance travelled in one second on a surface with $a = 2.5$ Å would be about 0.4 μm.

For a chemisorbed molecule the migration energy may be much lower, thus, many molecular adsorbates have a high mobility even at room temperature.

For a physisorbed molecule, for which E_m can be as low as 1 kcal/mol, f would be 4×10^{11} at 100 K on the same surface, corresponding to a distance of 0.2 mm.

5 SURFACE COMPOSITION

The chemical, electronic and mechanical properties of solids critically depend on the composition of the surface. The performances of a catalyst consisting of supported small bimetallic particles and the characteristics of a protective coating are determined by the composition of the topmost atomic layer. Therefore, it is of great relevance to understand and control the physical–chemical parameters which affect the surface composition, as well as to determine the composition of the few top-most atomic layers by means of suitable surface techniques. In the first part of this section, the thermodynamics guidelines governing the changes in surface composition will be reviewed. Minimisation of the surface free energy of the condensed phase system is the driving force for surface segregation of bulk impurity and for changes of composition of alloys and other multicomponent systems. In the second part of this section, the general principles and applications of the most common techniques used to determine surface composition will be reported. In the last part, some examples of surface segregation for alloy surfaces and of variation of surface composition of multicomponent systems caused by gas–solid interaction or by physical treatments will be discussed.

5.1   THERMODYNAMICS OF SURFACES

Let us consider those aspects of surface thermodynamics which are relevant in determining the surface composition. The surface contribution to the free energy G of a solid is always positive. In order to create a surface work must be done on the system which involves the breaking of bonds at the surface. The work required to increase the surface area by dA in a reversible process at constant temperature and pressure is related to the surface tension of a solid since

$$dW_{p,T} = \gamma dA \qquad (2.21)$$

where γ is the surface tension of the condensed phase.

For a one-component system the following relationship exists between γ and the surface free energy G^S (expressed as energy per unit area)

$$\gamma = G^S + A\left(\frac{\partial G^S}{\partial A}\right)_{p,T} \qquad (2.22)$$

The two terms on the right side of equation (2.22) distinguish between the work done to form a new surface dA and the work done in stretching an existing surface to increase the surface area by dA. In the last case the number of surface atoms is fixed while the state of strain of the surface is changed.

If the surface is unstrained so that the second term on the right-hand side of equation (2.22) is equal to 0, the surface tension is equal to the surface free energy

$$\gamma = G^S = \left(\frac{\partial G}{\partial A}\right)_{p,T} \qquad (2.23)$$

For a multicomponent system, from the relationships expressing the change of the surface free energy as a function of composition, surface tension and temperature, the Gibbs equation can be derived:

$$d\gamma = -S^S dT - \sum_i \Gamma_i d\mu_i \qquad (2.24)$$

where S^S is the surface entropy; μ_i is the chemical potential of species i; $\Gamma_i = n_i^S/A$ is the surface excess of component i, n_i^S being the difference between the number of moles actually present at the surface and the number of moles that would be present if the system were homogeneous up to the surface. For ideal systems or for dilute real systems exhibiting ideal behaviour, the chemical potential μ_i is related to the mole fraction x_i by the equation

$$\mu_i = \mu_i^0 + RT \ln x_i \tag{2.25}$$

where μ_i^0 is the standard chemical potential of the pure component i.

At constant temperature, substituting the expression for the chemical potential (2.25), equation (2.24) becomes

$$\Gamma_i = -\frac{1}{RT}\left(\frac{\partial \gamma}{\partial \ln x_i}\right)_{p,T,nj \neq i} \tag{2.26}$$

On the basis of this equation, the surface concentration of one component (expressed as surface excess) can be determined from the dependence of the surface tension upon the bulk concentration of that component. The Gibbs equation indicates that, if the surface tension decreases upon increasing the bulk concentration of component i, this component will accumulate at the surface, i.e. $\Gamma_i > 0$.

Therefore, if the bulk contains impurities which lower the surface tension (for instance, sulphur or carbon), the surface of the solid will be covered by a layer of each impurities. In other words, bulk impurities acting as surfactants concentrate at the surface if the temperature is high enough to allow their diffusion. An impurity concentration of the order of a few parts per million may be sufficient to form a complete monolayer at the surface: a closed packed plane of atoms contains about 10^{15} atoms/cm^2, while one part per million in the bulk corresponds to about 10^{16} atoms/cm^3.

Let us consider the surface composition of a bimetallic alloy $A_xB_{(1-x)}$. In order to derive the relationship between surface and bulk composition, ideal solution behaviour is assumed. This means that the heat of mixing of the alloy is zero. It is also assumed that the A and B atoms have the same size. Moreover, we consider that only the composition of the topmost layer differs from the bulk composition. The following equation can be derived by equating the chemical potential in the surface and in the bulk

$$\frac{X_A^S}{X_B^S} = \frac{X_A}{X_B}\exp\left(-\frac{(\gamma_A - \gamma_B)a}{RT}\right) \tag{2.27}$$

where X_A^S and X_A are the atomic fractions of species A at the surface and in the bulk, respectively; the same for species B; a is the area occupied by one mole of the component; γ_A and γ_B are the surface tensions of pure A and B. From this equation it can be seen that the component with the lower surface tension will accumulate at the surface.

Due to the inherent difficulty of measuring surface tensions for solids, equation (2.26) and (2.27) are of limited value in predicting the composition of the surface region. In order to calculate surface composition, theoretical treatments based on statistical thermodynamics are needed. Most of these theoretical treatments use short-range pair potentials to simulate interaction energies in alloys. Model calculations should consider that metallic alloys are not ideal solutions since they have some finite heat of mixing. Metallic alloys can be described in the framework of regular solution approximation (that is, ΔH_{mix} is not 0 but the entropy of mixing is the same as an ideal solution) (see references [6] and [12]). A statistical thermodynamics treatment based on the regular solution approximation leads to the following relationship between the bulk composition and the composition of the topmost layer:

$$\frac{X_A^S}{X_B^S} = \frac{X_A}{X_B} \exp\left[-\frac{(\phi_{BB} - \phi_{AA})}{2kT}(z - z^s)\right] + \frac{1}{2}\Delta E[(z - z_n^S)(1 - 2X_A) - z_t^S(1 - 2X_A^S)]$$

(2.28)

where z is the bulk coordination number; z_n^S is the coordination number of atoms in the outermost layer; z_t^S is the number of bonds between atoms in the outermost layer and atoms in the second layer; z^S is the coordination number of the surface atoms which is equal to the sum of z_n^S and z_t^S; ϕ_{AA} and ϕ_{BB} are interaction energies corresponding to pairs AA and BB. ΔE is defined as

$$\Delta E = \phi_{AB} - \frac{1}{2}(\phi_{AA} + \phi_{BB})$$

(2.29)

where ϕ_{AB} is the interaction energy of pairs AB.

These interaction energies can be calculated using quantum mechanics (with various levels of approximation) or derived from thermochemical data. The interaction energies between like atoms can be obtained from the sublimation enthalpies, since

$$\Delta H_{sub,A} = -\frac{1}{2}z\phi_{AA}N_{Av}$$

(2.30)

where z is the bulk coordination number and N_{Av} is Avogardo's number.

The term ϕ_{AB} can be derived from the heat of mixing of the alloy.

For ideal solutions the interaction energy between unlike atoms ϕ_{AB} is equal to the average of the interaction energy between like atoms

$$\phi_{AB} = \frac{1}{2}(\phi_{AA} + \phi_{BB})$$

(2.31)

Hence, on the basis of equation (2.31) the term ΔE is 0. If we set $\Delta E = 0$ in equation (2.28) we obtain equation (2.27), derived for surface segregation in ideal systems, since

$$a\gamma_A = -\frac{1}{2}(z - z^s)\phi_{AA}N_{Av}$$

(2.32)

where a is the area occupied by one mole of A.

An additional contribution to surface segregation comes from reduction of strain in the lattice due to the different atomic size of the species constituting the alloy. Accumulation at the surface of one component can reduce the strain energy. A term which includes the difference between the strain energy produced by locating atoms of different size in the bulk and in the surface is introduced in equation (2.28) to take this effect into account. The theoretical treatment of surface segregation described above can be refined taking into account that several layers below the surface, and not only the topmost layer, may have a composition different from the bulk. The equilibrium surface composition is established relatively slowly, since the exchange of atoms between bulk and surface is diffusion limited. Surface composition not in equilibrium with the bulk can be obtained by cleavage of a solid at room temperature because of the low diffusion rates. Therefore, the possibility of a metastable surface composition should be considered when comparing experimental data with equilibrium surface composition derived from model calculations.

For more details on surface thermodynamics and equilibrium composition, see references [6] and [12].)

5.2 DETERMINATION OF THE SURFACE COMPOSITION

The most common techniques for surface analysis use low energy electrons or ions as a probe. As discussed previously, the strong interaction of low kinetic energy electrons with the matter allows one to obtain a signal which comes from the outermost layers of a solid. Techniques based on low energy ions are virtually sensitive to the composition of the topmost layer only.

Auger electron spectroscopy (AES) and X-ray photoelectron spectroscopy (XPS) are based on the energy analysis of electrons emitted from a surface upon bombardment with a suitable radiation. In the following part, the basic principles and applications to the surface chemical analysis of AES and XPS are described.

A core level of an atom can be ionised by electrons or photons of high enough energy, impinging on the sample. In the subsequent de-excitation process, the core hole can be filled by an electron which decays from a less tightly bound level. The energy released in this process can be emitted in the form of an X-ray photon (emission of characteristic X-rays) or used to emit a second electron. Electrons emitted in the second case are called Auger electrons, from the name of the scientist, P. Auger, who discovered the phenomenon. The kinetic energy of an Auger electron for the transition schematised in Figure 2.14 is given by

$$E_{ABC} = E_A - E_B - E_C \qquad (2.33)$$

where $E_A - E_B$ is the energy released by an electron falling from the B level to the core hole in the A level and E_C is the binding energy of the electron in the C level. After Auger emission, the atom is left in a doubly ionised state. As we can see from equation (2.33), the kinetic energy of an Auger electron is independent of the energy of the electron or X-ray photon that caused the core hole, but it is deter-

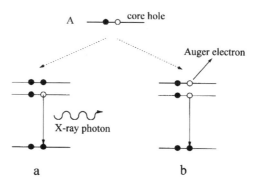

Figure 2.14. Schematic representation of de-excitation processes of atomic core holes: (a) Emission of X-ray photon. (b) Emission of an Auger electron.

mined by the energies of the levels involved in the transition. When the energy $(E_A - E_B)$ is below 1000 eV the Auger emission is more probable than X-ray emission. Therefore, the most intense Auger transitions for the various elements correspond to relatively low kinetic energies. The Auger transitions are indicated using the notation for the electronic levels of the X-ray spectroscopy. Thus, a $KL_1L_{2,3}$ Auger transition involves a core hole in the K shell, a decay of an electron from the L_1 level and the emission of an electron from the $L_{2,3}$ subshell.

AES is typically performed using a beam of electrons of energy ranging from 2 to 5 keV to excite the Auger emission. The electron energy analysers used in AES are the retarding field analyser and the cylindrical mirror analyser. The retarding field analyser uses the same electron optics as LEED. The energy distribution curve is collected by varying the potential applied to the retarding grid, which acts as an energy filter. The cylindrical mirror analyser (CMA) consists of two coaxial cylinders. Applying a potential ramp between the cylinders it is possible to select in energy the electrons emitted from the sample (positioned on the cylinder axis at a given distance from their entrance) which reach the detector (also positioned on the axis at the opposite side). Auger spectra are measured as first derivative of the electron distribution curve $N(E)$ with respect to energy, using modulation techniques and lock-in amplifiers. In this way the Auger signal can be separated from the continuous background due to the secondary electrons and inelastically scattered electrons. A typical Auger spectrum is shown in Figure 2.15.

By suitable analysis of the experimental data and using appropriate standards or tabulated sensitivity factors, AES can provide quantitative chemical analysis in addition to elemental compositional analysis of the surface. However, quantitative analysis is more problematic than in XPS. The sensitivity of AES is about 1% of a

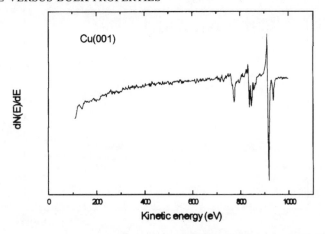

Figure 2.15. Auger spectrum of an ion bombarded Cu surface, excited with 3 keV electrons. As explained in the text, electron-induced Auger spectra are measured as the first derivative of the electron distribution curve with respect to energy. The LMM group of transitions (observed for copper in the range 700–1000 eV) is characteristic of 3d metals. The absence of any O or C signals (around 510 and 270 eV, respectively) indicates that surface is clean, as a result of the Ar^+ ion bombardment, which removed the initial impurities.

monolayer. Chemical state information is less easily obtained with AES than with XPS. However, when one of the electronic levels involved in the Auger transition is a valence level, different peak shapes are observed depending upon the chemical state of the emitting atom. The main advantages of AES with respect to XPS and other techniques for surface analysis is the high lateral spatial resolution (about 500 Å) which can be achieved by focusing the primary electron beam, allowing a map of the elemental distribution to be obtained. Another advantage of AES with respect to XPS is the higher surface sensitivity because relatively low kinetic energy electrons can be used (50–100 eV). When using AES, however, a possible damage of the surface induced by the primary electron beam should be taken into account, particularly when working with non-metallic solids and/or molecular adsorbates.

X-ray photoelectron spectroscopy (XPS), also called electron spectroscopy for chemical analysis (ESCA), is based on photoemission (see Section 3.3). Using monochromatic X-ray photons, it is possible to eject electrons from the core levels of atoms. The kinetic energy of the emitted electrons is related to the binding energy of a core level by equation (2.10), reported in the section dealing with UPS. In an XPS spectrum, peaks are present which correspond to emission of electrons from the various atomic core levels. A typical XPS spectrum is shown in Figure 2.16. In addition to photoemission peaks, in an XPS spectrum peaks are also observed due to Auger emission which is the consequence of the core level ioni-

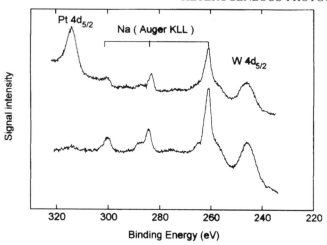

Figure 2.16. XPS spectrum of the (100) face of a 'tungsten bronze' single crystal, with composition $Na_{0.7}WO_4$, before (lower curve) and after (upper curve) deposition of three monolayers of platinum. The spectrum was obtained using Mg $K\alpha$ radiation (1253.6 eV). Besides the 4d XPS peaks of W and Pt, the KLL Auger emissions of Na (excited by the X-ray photons) are observed in the same region.

sation. On the low kinetic energy (high binding energy) side of a photoemission peak other features can be observed. Satellite peaks a few electronvolts lower in kinetic energy correspond to photoemission processes in which the ion is left in electronically excited states (shake-up peaks). Other features are due to photo-electrons which have lost discrete amounts of energy to excite plasmons of the solid (a plasmon is a collective oscillation of the valence electrons).

The most common X-ray sources used to produce XPS spectra employ Al ($K\alpha$ emission at 1486.6 eV) and Mg ($K\alpha$ emission at 1253.6 eV) anodes. The X-ray source can be equipped with a monochromator to reduce the line width characteristic of the emitted X-rays. Typically, the photoemitted electrons are energy analysed using a concentric hemispherical analyser (CHA). In this kind of analyser, in order to select the incoming electrons according to their energy, either a voltage ramp is applied to the two hemispheres or the voltage is fixed at a given value and a ramp (applied to a first stage of electron optics that has the additional function of focusing the photoelectrons on the entrance slit) allows retardation of the electrons at the pass energy. The latter method is preferred, since the instrument resolution is constant for the whole kinetic energy range of the photoelectrons. Hemispherical analysers provide the high resolution and the accuracy (typically 0.1 eV) in the determination of the kinetic energies required in XPS. Moreover, it is possible to perform angle-resolved measurements, since the fraction of emission solid angle that some hemispherical analysers accept is small enough (see XPD, Section 2.2.2).

From the area of the XPS peaks, obtained after subtraction of the inelastic scattered electron background, and using reference standards or tabulated sensitivity factors, a quantitative surface analysis can be performed (see also reference [13]).

The intensity of photoelectrons emitted in a given peak from atomic species i, whose concentration is constant over the depth explored, is given by

$$I_i = I_0 T(E) A n_i \left(\frac{d\sigma}{d\Omega}\right)_\alpha \Omega \lambda \cos \theta \tag{2.34}$$

where I_0 is the intensity of the primary photons; $T(E)$ is an instrumental transmittance function, varying with the kinetic energy E of the photoelectrons; A is the sample area explored; n_i the concentration of the emitting species, in atoms/unit volume; $(d\sigma/d\Omega)_\alpha$ is the differential cross-section for the photoemission process and is a function of the angle α at which the photoelectrons are collected with respect to the direction of the incoming photons; Ω is the solid angle of acceptance of the spectrometer; λ is the electron mean free path, that is, the average distance an electron can travel in the solid with a probability $1/e$ of no energy loss; θ is the angle at which the photoelectrons are collected with respect to the surface normal.

The above equation allows, at least in principle, the concentration of a given atomic species to be determined, if all the physical quantities are known. Actually, the easiest way is to calibrate the instrumental sensitivity to each atomic species using samples with known composition. A difficulty arises, however, because of the parameter λ which not only varies as a function of the kinetic energy but also with the nature of the solid matrix; strictly speaking, this prevents comparison of the signals of the same species present in two different solids, unless the values of the λ parameter can be evaluated accurately. Now there are several more or less empirical methods, allowing an evaluation of λ suitable for most purposes. In conclusion, XPS can be used for a quantitative analysis of a solid surface, at least in the case in which the composition does not vary within the depth explored. The depth below the surface from which 95% of the XPS signal comes, at normal emission, is 3λ, while the depth contributing 50% of the signal is 0.7λ. Since λ may vary between 10 to 20 Å, the explored depth in XPS is about 30–50 Å, with the first 7–15 Å contributing half the intensity. These values show how electron spectroscopy is sensitive to the surface composition.

The quantitative determination of the surface composition is more complicated if the concentration of atomic species varies within the explored depth. A relatively simple case is that of a thin layer of thickness d with constant composition, covering the solid. In this case, the signal of an atomic species present in the layer is given by

$$I_i^S = I_0 T(E) A n_i^S \left(\frac{d\sigma}{d\Omega}\right)_\alpha \Omega \lambda^S \cos \theta \left[1 - \exp\left(-\frac{d}{\lambda^S \cos \theta}\right)\right] \tag{2.35}$$

where the index S is relative to the layer. This equation allows determination of the thickness d if n_i^S is known and, again, the other quantities can be calibrated on a solid having a λ not too different from that of the layer. A typical case is that of thin oxide layers, for which the parameters are easily determined on the bulk oxide.

If the same element is also present in the solid covered by the thin layer, two cases can be found: (i) the XPS signal coming from the atoms of the layer cannot be separated from that coming from the bulk; (ii) the signal coming from the layer has a binding energy sufficiently different to be separable from the bulk signal. The first unfavourable case is relatively rare, since XPS is a very sensitive to the bonding situation of an atomic species, as will be discussed below; usually the second case is found in XPS experiments. Thus, the signal coming from the atoms of the solid covered from the thin layer will be

$$I_i^B = I_0 T(E) A n_i^B \left(\frac{d\sigma}{d\Omega}\right)_\alpha \Omega \lambda^B \cos\theta \exp\left(-\frac{d}{\lambda^S \cos\theta}\right) \qquad (2.36)$$

where the index B now refers to the bulk solid, the exponential term representing the attenuation of the signal by the layer. In this case, it is possible to determine directly the ratio I_i^S/I_i^B in the same spectrum, and to derive the thickness d of the layer without the need of other calibrations, since most of the parameters will vanish in the ratio between equations (2.35) and (2.36). Of course, the compositions of the two solid phases, or at least their ratio, must be known, as must be known the two parameters λ^S and λ^B.

The above treatment can be applied only to perfectly flat surfaces. In many cases, the sample surface is corrugated on a microscopic scale, and the thickness of a surface layer can be only roughly estimated.

For the case of a composition varying continuously from the surface to the bulk, in general the variation law cannot be determined, but a model profile can describe the XPS data better than another, particularly if several XPS spectra are obtained at different emerging angles of the photoelectrons.

The sensitivity of XPS is around a few percent of a monolayer, but it strongly depends on the element and on the particular level excited. An advantage of XPS with respect to AES is the possibility of studying molecular adsorbates and surfaces of organic materials since the damage caused by X-ray irradiation is much more limited than in the case of electron bombardment.

But the most relevant feature of XPS is the possibility of determining the chemical state of elements at the surface. The binding energies of the core levels shift as a result of changes in the formal oxidation state, molecular environment, difference in lattice site and so on. These shifts of the binding energies are due to changes of the charge density in the valence levels involved in the chemical bonds and they are called chemical shifts. Without entering into the details of a rigorous

treatment, the chemical shift for an atom in two different bonding environments can
be roughly derived by the so called ground-state potential model:

$$\Delta E_i = E_i^{(2)} - E_i^{(1)} = k(q_i^{(2)} - q_i^{(1)}) + (V_i^{(2)} - V_i^{(1)}) \qquad (2.37)$$

where

$$V_i = \Sigma_{j \neq i} \frac{q_j}{r_{ij}}$$

In equation (2.37), $E_i^{(1)}$ and $E_i^{(2)}$ (in atomic units) are the binding energies of a core
level of atom i in the two different chemical states 1 and 2, k is a proportionality
constant, $q_i^{(1)}$ and $q_i^{(2)}$ are the effective changes in the valence levels for atom i in
the two chemical environments (expressed as fraction of elementary charge). V_i is
the contribution to the potential energy of an electron at the nucleus of atom i due to
the point charges q_j on the surrounding atoms at r_{ij}. The constant k can be roughly
estimated theoretically as the Coulomb integral between an electron in a valence
orbital and one in the core orbital. In fact, it is usually derived by fitting the
observed chemical shifts of a given atomic species in several compounds using the
charges calculated with an HF-SCF method. The V_i terms in equation (2.37) tend to
counterbalance the effect caused by the valence charge on the central atom.
However, the latter usually prevails, so that a loss of negative charge (oxidation) is
generally accompanied by an increase of the binding energies of core levels (see
Figure 2.17 and Table 2.4).

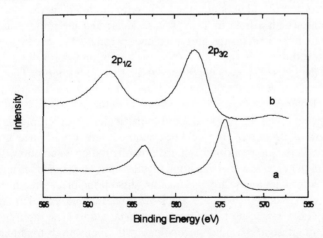

Figure 2.17. Example of chemical shift observed for chromium deposited on ZnO: (a) the
2p doublet of about ten monolayers of Cr deposited in ultra-high vacuum on the (0001) face
of ZnO at room temperature; (b) after annealing at 550° C, the Cr atoms are embedded in the
oxide matrix. Mg Kα excitation source.

Table 2.4. Binding energies of C 1s level in various solid compounds (from reference [21])

Carbon compound	C1s binding energy (eV)
Graphite	284.5
Fe_3C	283.9
TiC	281.6
$Fe(CO)_5$	288
Na_2CO_3	289.4
C_6H_6	284.7
C_6F_6	289.5

Equation (2.37) does not take into account final state effects (like a relaxation of the valence electrons) which may significantly contribute to the chemical shift. In particular, equation (2.37) is not valid for molecules in a condensed phase, where the polarisation of the surrounding molecules must also be considered. When working with insulators, an apparent increase of the binding energies may be observed, which can be attributed to a positive charging of the sample caused by the emission of electrons. In this case, charge neutralisation techniques (for instance, bombardment of the sample with very low-energy electrons) must be used in order to derive correct chemical state information from XPS, unless an element with exactly known binding energy is present at the surface. The XPS chemical shifts can be used empirically, since the binding energies for each element in several compounds are tabulated (see reference [13]). However, theoretical treatments are now available allowing reliable calculation of the binding energies for atoms in gaseous molecules, since the difference in total energy between the ionised molecule (with a hole in a given core level) and the ground state of the neutral molecule directly gives the ionisation energy (or electron binding energy) of that level. More difficult is the calculation for atoms in a solid, because all the atoms around the ionised species contribute to the final state energy. Thus a cluster including several neighbours around the ionised atom must be used, and the variation of the cluster properties as a function of the cluster size must be explored. But for solids containing heavy metals, and in particular transition metals, this is a formidable task if *ab initio* Hartree–Fock methods are used, since correlation effects become very important and require a configuration interaction (CI) treatment, which, in the case of a few transition metal atoms, soon becomes untreatable, unless heavy approximations are used. Now, first principle methods based on density functional theory (DFT) are available, and reliable calculations can be performed on clusters with up to ten transition metal atoms. The rapid increase of performances of present computer technology will soon allow treatment of big metal clusters simulating different crystal faces. This will permit complete exploitation of the information contained in XPS spectra, particularly in the case of molecules adsorbed on single crystal faces of known structure.

Both AES and XPS probe a limited depth below the surface which is determined by the mean free path of the electrons, 5–20 Å depending on the electron kinetic energy. But the ultimate surface sensitivity is obtained by means of low-energy ion scattering spectroscopy (LEIS) [14].

In LEIS, a monoenergetic beam of ions of a noble gas with low energy (typically of the order of 1 keV) is focused on the sample and the ions backscattered from the surface are energy analysed. In the collision process, a fraction of the primary energy is transferred from the incident ions to the target atoms on the sample, as in the case of an elastic collision between balls. From the treatment of the elastic collision between incident ions and target atoms, based on the conservation of energy and momentum, the ratio of the energy of the scattered ions E_1 and the energy of the incident ions E_0 can be derived:

$$\frac{E_1}{E_0} = \left[\frac{(M_2^2 - M_1^2 \sin^2 \theta)^{1/2} + M_1 \cos \theta}{M_1 + M_2} \right]^2 \qquad (2.38)$$

where M_1 and M_2 are the masses of the incident ion and of the target atom, respectively and θ is the scattering angle defined by the incident beam direction and the axis of the analyser. This equation is derived for the more common case $M_1 < M_2$, for which there are no limits for the scattering angles. Helium ions are the most commonly used as a probe, since they allow all target masses (excluding hydrogen) to be detected.

Thus, from the LEIS spectrum and using equation (2.38) it is possible to determine the elements present on the surface of the sample.

Since at low kinetic energy the noble gas ions penetrating beyond the first atomic plane of the surface have a high probability of being neutralised, the scattered ions reaching the analyser carry information only about the composition of the topmost layer. As an example, Figure 2.18 shows the case of the (100) surface of a single crystal of a 'tungsten bronze', with composition $Na_{0.7}WO_4$, whose XPS spectrum was reported in Figure 2.16. For this solid, two possible terminations of the surface are possible and the LEIS spectrum clearly shows that the actual termination is the NaO plane, since no W signal is observed. Another example is the (100) face of a Pt_3Ti alloy, an ordered alloy with the Cu_3Au-type structure. As shown in Figure 2.20, two terminations are possible for the (100) face, and the LEIS spectrum of the clean surface (see Figure 2.21) is in agreement with the pure Pt termination.

The sensitivity of LEIS varies widely depending on the masses of the probe and target atoms (in general, for a given probe it increases with increasing target mass). In favourable cases, it can be around 1% of a monolayer. In addition to elemental composition, LEIS can provide quantitative chemical analysis of the topmost layer of a solid. However, a quantitative analysis is possible only if target atomic species with relatively close masses are involved, or if the scattering cross sections for the target species can be calculated. But the main difficulty arises from the neutralisation probability, which affects the peak intensities and, for a given element, depends on the

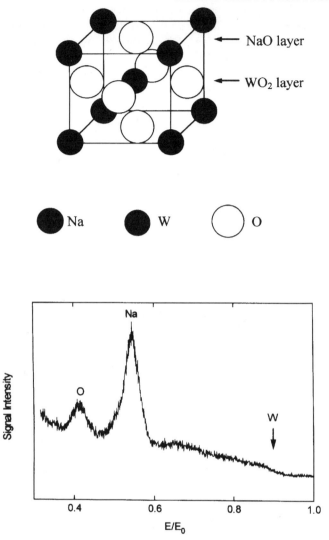

Figure 2.18. LEIS spectrum of the (100) face of a 'tungsten bronze' single crystal, with composition $Na_{0.7}WO_4$ (same as that of Figure 2.16) obtained with 600 eV He$^+$ ions. The upper part of the figure shows the two possible terminations of the surface.

solid matrix. Thus, an empirical calibration of the relative elemental sensitivities obtained for different solids may not be reliable, since they may vary depending on the solid matrix. However, it has been found that, for a given matrix with variable composition (like a solid solution), the relative sensitivities of two elements are conserved.

No chemical state information can be derived from LEIS spectra, since the peak position an atomic species depends only on its mass. A possible surface damage due to the incident ion beam should be considered when using LEIS, although employing low-mass probing atoms, in particular helium, the damage can be reduced to a negligible level.

A technique based on the damage produced by an energetic ion beam (10–30 keV) hitting the surface of a solid is secondary ion mass spectroscopy (SIMS). A fraction of the material sputtered from the surface of the sample under ion bombardment can be emitted as ionic species. In SIMS the masses of the secondary ions emitted from the surface of the sample are analysed using a mass spectrometer. The sensitivity of SIMS can reach 10^{-6} monolayer. Therefore, the most common application of SIMS is the determination of the composition profile of low concentration impurities in solids. Another relevant feature of SIMS is its ability to detect hydrogen.

Unfortunately, quantification of SIMS data is problematic due to the complexity of the sputtering process which produces ions both positive and negative, singly and multiply charged, neutrals and cluster of atoms. The cross-sections of the different events cannot be easily quantified.

5.3   SURFACE VS. BULK COMPOSITION

Surface segregation, although it is a general phenomenon, has been investigated in particular for alloys. A large amount of data has been collected for the composition of surfaces of single crystals of random substitutional alloys. In these alloys the lattice sites are randomly occupied by atoms A or B with a probability weighted by the atomic fraction of the species constituting the alloy. The results of model calculations predict correctly the surface composition, the composition profile and the segregation dependence upon bulk concentration and the crystallographic orientation of the surface for this class of alloys. From the comparison of experimental and calculated data some general trends can be derived. The components with the lower surface tension and the larger atomic radius tend to segregate at the surface of alloys. Changes in the composition involve not only the topmost layer but the first 3–4 layers below the surface. Damped oscillatory compositional profiles have been found at the surfaces of Pt–Ni and Pt–Co alloys (see Figure 2.19).

The variation in the composition of the surface region with respect to the bulk may induce changes in the structure of the surface region, such as a relaxation of the interplanar distances and, as observed in some cases, a reconstruction of the surface. For instance, a hexagonal reconstruction of the topmost layer of pure Pt has been observed for the (001) surface of the $Pt_{80}Co_{20}$ alloy (references [15] and [16]).

In the case of ordered substitutional alloys, characterised by a relatively high heat of mixing, surface segregation is hampered by the fact that bonds between unlike atoms must be broken. However, for this kind of alloys and other compounds, where different terminations of the bulk structure are possible, the topmost

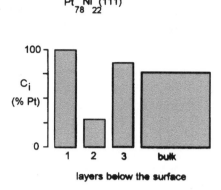

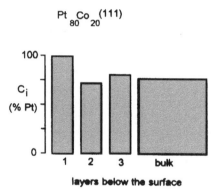

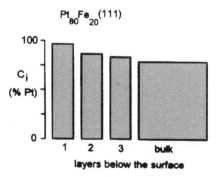

Figure 2.19.   Compositional profiles for the (111) surface of random substitutional alloys of platinum. Pt–Ni and Pt–Co show a damped oscillatory enrichment of Pt at the surface. Pt–Fe alloy shows a damped monotonous enrichment (Data taken from reference [20]).

layer can have a different composition relative to the bulk one. For instance, let us consider substitutional alloys A_3B with a Cu_3Au-type structure. This type of alloy has an fcc structure and the A atoms at the corners of the centred unit cell are substituted by B atoms (Figure 2.20). For example, Pt_3Sn, Pt_3Ti and Ni_3Al alloys have this kind of structure. For the (100) and (110) surfaces of this group of alloys two terminations of the bulk structure are possible: one termination layer consists of pure A and the other termination layer contains 50% atomic fraction of A and 50% atomic fraction of B. The stacking of these alternating layers reproduce the A_3B composition of the alloy. The (100) and the (110) surfaces of the Pt_3Sn alloy terminate with the mixed planes containing 50% Pt and 50% Sn. Hence, the topmost layer of the alloy is enriched in Sn with respect to the average composition of the bulk [17]. On the other hand, the (100) surface of the Pt_3Ti alloy terminates with a layer of pure Pt, as demonstrated by LEIS (Figure 2.21) and LEED [18].

The surface composition of a solid can be altered by means of various treatments. A simple one consists in annealing the sample at high temperatures in vacuum. This is an effective way to change the surface composition of oxide or compounds which have components with relatively high vapour pressure. Some oxide surfaces can lose oxygen upon heating in vacuum. The loss of oxygen pro-

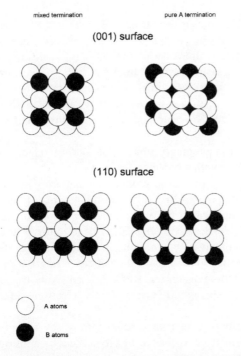

Figure 2.20. The two different terminations for the (001) and (110) surfaces of ordered substitutional alloys with Cu_3Au-type structure.

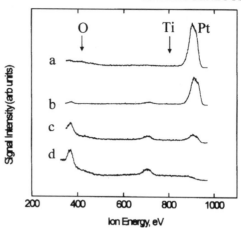

Figure 2.21. LEIS spectra (He$^+$, 1 keV) of the (100) face of a Pt$_3$Ti alloy (see structure in Figure 2.20), upon gradual oxidation by exposure to 10^{-6} torr oxygen at about 700°C: (a) Clean surface annealed at 800°C (only Pt in the first atomic plane); (b)–(d) gradual formation of a thin layer of a TiO-like phase covering the alloy surface. (U. Bardi and G. Rovida, unpublished results).

duces a non-stoichiometric surface phase with lower oxidation state of the element and oxygen vacancies. If the vacancies are ordered, superstructures can be observed. For instance in the case of the TiO$_2$(100) surface (1 × 3), (1 × 5) and (1 × 7) unit cells have been observed upon annealing in vacuum. Reconstructions induced by change in the surface stoichiometry have been found also in the case of the Al$_2$O$_3$(0001) surface and for the V$_2$O$_5$(010) surface. In some cases, the equilibrium composition can be restored upon heating the modified oxide surface in oxygen atmosphere.

Changes in the surface composition of the surface region can be obtained by bombarding the surface with a noble gas ion beam or with an electron beam. An ion beam sent on the surface of a solid produces a sputtering of the surface atoms. If one species is sputtered more efficiently than the others, the surface region will be depleted in that component. The preferential sputtering of one component can substantially modify the surface composition of an alloy over a depth of several layers. Equilibrium composition can be restored by annealing the alloy. For oxide surfaces, the preferential sputtering of oxygen can produce a reduced oxide layer. An electron beam impinging on the surface of an oxide can also alter the stoichiometry of the surface region, because of the electron stimulated emission of oxygen atoms.

Surface composition can of course be varied by solid–gas interactions. As an example, we will consider the change in the surface composition due to the oxidation of alloy surfaces. If one component of the alloy has a greater affinity for oxygen than the other, as in the case of Pt–Ti or Pt–Co alloys, an oxide layer of the more reactive species forms at the surface of the alloy upon exposure to oxygen. Figure 2.21 shows the gradual oxidation of the (100) face of a Pt_3Ti single crystal exposed to oxygen at temperatures high enough to allow Ti atoms to diffuse from the bulk to the surface and form a thin layer of a TiO-like phase, which gradually covers the alloy surface. If the oxidation is carried out at relatively low temperatures at which diffusion is a slow process, at the oxide–metal interface the alloy will be depleted in the component which forms the oxide layer.

So far we have considered segregation at ideally flat surfaces of single crystals. Surface defects, such as steps, kinks, dislocations and grain boundaries play a role in surface segregation. Since structural irregularities are sites of lower coordination, defects are usually more enriched in the segregating species than are sites on flat terraces.

REFERENCES

[1] D. P. Woodruff and T. A. Delchar, *Modern Techniques of Surface Science*, second edition, Cambridge University Press, Cambridge, 1994.
[2] E. A. Wood, *J. Appl. Phys.* **35** (1964), 1306.
[3] J. B. Pendry, *Low Energy Electron Diffraction*, Academic Press, New York, 1974.
[4] M. A. Van Hove, W. H. Weinberg and C. M. Chan, *Low- Energy Electron Diffraction*, Springer-Verlag, Berlin, 1986.
[5] A. Atrei, U. Bardi, J. X. Wu, E. Zanazzi and G. Rovida, *Surface Sci.* **290** (1993), 286.
[6] G. A. Somorjai, *Introduction to Surface Chemistry*, Wiley, New York, 1994.
[7] P. R. Watson, M. A. Van Hove and K. Hermann, Atlas of surface structures, *Journal of Physical and Chemical Reference Data*, Monograph N.5.
[8] S. D. Kevan, *Physical Review Letters* **50** (1983), 526.
[9] G. Ertl and J. Kuppers, *Low Energy Electrons and Surface Chemistry*, VCH, Weinheim, 1985.
[10] J. E. Inglesfield and B. W. Holland, in *The Chemical Physics of Solid Surfaces and Heterogeneous Catalysis*, Vol. 1, Chapter 3. D. A. King and D. P. Woodruff, Eds., Elsevier, Amsterdam, 1981.
[11] A. Zangwill, *Physics at Surfaces*, Cambridge University Press, Cambridge, 1989.
[12] M. J. Spaarnay, *Surface Science Reports* **4** (1985), 101.
[13] D. Briggs and M. P. Seah, *Practical Surface Analysis*, Wiley, New York, 1985.
[14] L. C. Feldman and J. W. Mayer, *Fundamentals of Surface and Thin Film Analysis*, North-Holland, Amsterdam, 1986.
[15] U. Bardi, A. Atrei, P. N. Ross, E. Zanazzi and G. Rovida, *Surface Sci.* **211/212** (1989), 441.
[16] U. Bardi, A. Atrei, E. Zanazzi, G. Rovida and P. N. Ross, *Vacuum* **41** (1990), 437.
[17] A. Atrei, U. Bardi, G. Rovida, M. Torrini, E. Zanazzi and P. N. Ross, *Physical Rev. B* **46** (1992), 1649.
[18] A. Atrei, L. Pedocchi, U. Bardi, G. Rovida, M. Torrini, E. Zanazzi, M. A. Van Hove and P. N. Ross, *Surface Sci.* **261** (1992), 64.
[19] A. Atrei, U. Bardi, G. Rovida and E. Zanazzi, *Vacuum* (1990), 333.
[20] Y. Gauthier and R. Baudoing, in *Segregation and Related Phenomena*, P. Dowben and A. Miller, Eds., CRC Press, Boca Raton, FL, 1990.
[21] J. Chastain (ed.), *Handbook of X-ray Photoelectron Spectroscopy*, Perkin-Elmer, Norwalk, CT, 1992.

3 Photoadsorption and Photo-desorption at the Gas–Solid Interface

R. I. BICKLEY

University of Bradford, BD7 1DP, UK

1 INTRODUCTION

Photosorption is a phenomenon which relates the manner in which a flux of electromagnetic radiation interacts with one or more of the participants in an equilibrium process involving a gas and the surface of a solid. We may describe a generalised thermally activated adsorption equilibrium in the following terms:

$$X(g) + S \underset{k_d}{\overset{k_a}{\rightleftharpoons}} X - S(ads)$$

the equilibrium surface coverage, θ_x, being described by some form of adsorption isotherm (e.g., Langmuir) where k_a and k_d represent the rate constants for the adsorption and desorption steps. Clearly the shape of the isotherm for the interaction of $X(g)$ with the surface, S, will be dictated by the magnitude of the inter-

Heterogeneous Photocatalysis, Edited by M. Schiavello
© 1997 John Wiley & Sons Ltd.

action, the curvature of the function $\theta_x = f(p_x)_T$ being sharper as the magnitude of $-\Delta H(\text{ads})$ increases, where $-\Delta H(\text{ads})$ is the molar enthalpy change arising from the adsorption process, or in kinetic terms by the manner in which the thermal energy (kT) provided by the surroundings produce more favourable conditions for the adsorption step relative to the desorption step or vice versa.

Under these conditions of thermal equilibrium it is possible to irradiate the system with a flux of electromagnetic radiation which, in its broadest terms, can extend from radio-frequency and microwave radiation $(E_\lambda \approx 0.001 \text{ eV})$ through infra-red radiation $(E_\lambda \approx 0.01–0.1 \text{ eV})$, visible and near u.v. radiation, $(E_\lambda \approx 1–4 \text{ eV})$, to far u.v. $(E_\lambda \approx 20 \text{ eV})$, x-radiation $(E_\lambda \approx 1000 \text{ eV})$ and γ-radiation. It is to be expected therefore that the perturbations on the equilibrium imposed by each of these forms of radiation differ. The lower energy photons basically respond to the quantised vibrational and rotational energies which can exist either in the solid or in the adsorbed molecular species, $X(\text{ads})$ or in the gaseous molecules $X(\text{g})$, and by consequence can displace the position of equilibrium in a manner not dissimilar to that provided by a change of temperature in the surroundings. The major difference between displacements of equilibrium caused respectively by these sources will lie in their nature; namely the manner in which energy is distributed within the source of the perturbation, i.e., whether the radiation source is monochromatic or polychromatic, whether the radiation source is incoherent or coherent; and in the case of thermally perturbed systems the means by which the energy is transferred (e.g., conductive or radiative) but in either case the transfer process will be dictated by a Maxwell–Boltzmann distribution function (gaseous conduction) or an approximation to a black-body radiation function. Accordingly differences in the displacement of an equilibrium will emerge as direct consequences of the origins of these perturbations.

With larger forms of photonic energy transfer, the excitation of electrons from their ground states to higher states becomes possible during the absorption of the photon. Such excitations are possible with many crystalline solids in the near-u.v. region and in some cases in the visible region. The prerequisite for absorption of the radiation is that the transition probability, $W_{1 \to u}$, between the lower (l), occupied, electronic state and the upper (u), vacant, electronic state is non-zero.

$$W_{1 \to u} = \frac{2\pi}{\hbar} \int \Psi_u |h\nu| \Psi_1 \; d\tau \neq 0$$

In the context of the solid, itself, in isolation from a gas in equilibrium with its surface, Ψ_u and Ψ_1 can represent the empty conduction band and the almost full valence band of the bulk solid; or the surface states which are derived therefrom; or indeed electronic states which arise from intrinsic impurities or as a result of systematic additions of alio-valent materials (extrinsic impurities) and, by a corresponding argument, the influence that such materials can also exert resulting from their presence at surfaces.

In the presence of an adsorbing gas, further possibilities arise particularly where additional surface electronic states are created by the presence of the adsorbed moiety. The perturbations that are created in solids and in molecules by their exposure to X-radiation differ mainly in the following respects from those others which induce electronic transitions. Since the energies of the photons are many hundreds of times larger, then transitions are possible which could not occur with photons of lower energy. Transitions from the valence band of a solid to higher unoccupied bands are possible, photo-emission of electrons can occur and electrons can be excited from core electronic states with attendant relaxation processes such as Auger emission.

Finally, with γ-radiation, the effects of the solid being exposed to it are such that atomic/ionic displacements can occur with the creation of non-equilibrium populations of cationic and anionic vacancies which may be occupied by electrons or by positive holes; or it may indeed produce chemical transformations leading to the production of impurity centres in a manner analogous to that which arises from extrinsically introduced impurity species. With all forms of high-energy photons from the mid-u.v. range of ~ 6 eV to the upper limit of γ-radiation, the consequences of the interaction of these photons with gaseous molecules can result in the formation of free atoms in their electronic ground states, or in highly excited states ('hot' atoms), or even gaseous ions. It should not be surprising therefore that these forms of radiation produce effects which can modify the condition of the solid in its interaction with normal molecules, the interactions of excited molecules with the solid and/or the interactions between solids in their excited states and excited molecules. Clearly where these latter possibilities arise the chemical consequences are most complex and frequently lead to the introduction of a large degree of irreversibility of the overall chemical processes. They may be broadly classified as being surface photolytic processes and/or bulk photolytic processes. Such processes require high-energy photons in many cases, but the photographic process, involving silver halides, functions simply because a latent image of silver atoms can be created with visible light and the effectiveness of the response of the silver halide particles can be improved by the addition of impurity species which act as sensitisers.

Adsorption may be subdivided into two main classifications: physical adsorption (physisorption) and chemical adsorption (chemisorption). Physical adsorption is a manifestation of the existence of non-ideality in gases, in their interactions between other molecules, and in the interactions between individual molecules and the surfaces of solids to which they are exposed. In general adsorption interactions of this description become measurable only under conditions in which the mean kinetic energies of the gaseous molecules have been reduced to approximately the magnitudes of the interaction energies of the molecules with the surface or between two gaseous molecules. Thus physisorption becomes measurable below, or in the region of, the boiling point of the liquefied gas and relates to a distortion (polarisation) of the electron charge clouds to produce an overall attractive force, with a reduction of the Gibbs' free energy of the system $(-\Delta G)$, a reduction in entropy $(-\Delta S)$, in most cases, and the release of heat $(-\Delta H(\text{ads}))$. In general, however, this form of adsorption interaction is regarded as being fairly weak $(-\Delta H(\text{ads}) \leqslant 40 \text{ kJ mol}^{-1})$ and such processes are easily reversible. Thus small thermal changes arising from the absorption of electromagnetic radiation would be expected to create changes to the position of a gas–solid equilibrium of the form described previously. In such situations the increase of temperature of the system would lead to molecules being released from the adsorbed state into their gaseous form; desorption would have occurred. While the interactions between a particular gaseous molecule and a solid of a specific chemical composition may be defined as leading to physical adsorption, the physical morphology of the solid can create very substantial variations to the form of the equilibrium depending upon whether the solid is crystalline, porous (containing either micropores ($\emptyset < 5$ nm) or meso-pores ($\emptyset > 10$–20 nm)) or non- porous; massive or finely divided particulates or amorphous. When all the variations of these possibilities are considered, taken together with a comparison between the magnitude of the gaseous molecular interaction and that of the molecule with the surface under examination, six different isotherm shapes emerge, five of which were classified by Brunauer et al. in 1940 [1, 2] (The BDDT Classification) to which has been added the final, sixth, isotherm shape some twenty years ago. The isotherms have been classified as Type I, Type II etc. and the completion of the classification was achieved when it was realised that the Type I isotherm shape frequently measured did not arise for the reasons originally proposed by Irving Langmuir in 1916. The Langmuir isotherm illustrated the behaviour of an idealised surface in which the adsorption interaction energies were independent of the surface coverage. Such surfaces occur relatively infrequently and were reclassified as Type VI to differentiate them from similarly shaped Type I isotherms which are a manifestation of the existence of micropores in the solid (e.g., charcoal) [3]. Clearly the impact of electromagnetic radiation on such systems will be modified by the morphological variations and by the manners in which they influence the interaction energies of molecules with their surfaces.

    Chemisorption, on the other hand, is a strong interaction producing adsorption enthalpies $(-\Delta H_c)$ of the order of 100 kJ mol^{-1} or greater. In general the isotherm shape reflects the strength of the interaction, and it is frequently found to be of Type I. The major difference from physical adsorption is that the chemisorption process is activated in a manner analogous to a normal chemical reaction in that it possesses an activation energy of chemisorption. By implication it involves the creation of a strong chemical bond between the adsorbing molecule and the surface, the rate of formation increasing with increasing temperature in the chemical systems in which it is shown to arise, which accordingly further implies reaction specificity in contrast to the almost completely non-specific nature of physical adsorption. Chemisorption, as well as being specifically dependent between appropriately reactive partners i.e., gas molecule/adsorbate and the surface/adsorbent, is site-specific on the surface with the adsorbing molecule interacting with a limited number of atomic surface centres, which limits the maximum surface coverage to one monolayer, in contrast to the non-specific multi-layer adsorption which can occur with physical adsorption.

In such circumstances, the influence of electromagnetic radiation upon the equilibrium position of a chemisorption process will be almost entirely due to events which can either increase the extent of bonding via the activation of specific adsorption sites, or decrease it via the weakening/breaking of surface bonds which have been established already. Chemical bonding may be described broadly in terms of covalence, where the electronic contribution is provided jointly by the bonding partners; by ionic bonding in which electronic charge is transferred from one bonding partner to the other or by coordinate bonding, where one partner has the capacity to donate a lone pair of electrons into vacant orbitals possessed by the other partner. Accordingly, in circumstances where the absorption of electro-magnetic radiation by the solid or by the specific bond leads to a redistribution of electronic charge, a manifestation of this effect will be photosorption. Clearly, where the process favours an increase in the coverage of surface species, the phenomenon may be described as being *photoadsorption* while in the converse case the process may be described as *photodesorption*. Frequently, although the net shift of the equilibrium is in one direction, the use of isotopic labelling (e.g., $^{17}O_2$ or $^{18}O_2$) reveals that both processes (adsorption and desorption) may be affected, but that one process may be affected more extensively than the other rather than there being any absolute influence of the radiation upon one of the two opposing processes.

The majority of studies of heterogeneous photocatalysis have been undertaken using visible or near-u.v. radiation ($E_\lambda < 5$ eV). Such radiation sources simulate the most energetic solar radiation which penetrates the Earth's atmosphere and thus provides the impetus for laboratory studies of problems of terrestrial surface photochemistry. Some studies have been made using X-rays as the stimulus but these problems are somewhat esoteric and will not be addressed further within the present article.

Many of the solid materials (adsorbents) which are effective for photocatalysis exhibit their principal optical/electronic transitions in the near-u.v. region of the electromagnetic spectrum ($E_\lambda \approx 2.5$–5 eV). Two typical examples of such materials are zinc (II) oxide (ZnO) and titanium (IV) oxide (TiO_2), each possessing a fundamental electronic transition at approximately 3 eV, although a wide range of solids can be activated by photons within this range of energies [4, 5].

Sources of radiation suitable for producing photons of these energies are derived usually from passing electrical discharges through gases or vapours at low, medium or high pressures, e.g., Xe(g), Xe(g)–Hg(vap), or Hg(vap), these sources emitting their energies merely from the passage of excited atoms back into their ground electronic states in a random manner with stimulated emission contributing a minor proportion of radiant energy. In such circumstances many of the studies of photo-stimulated heterogeneous effects have been conducted with sources of this description. The advent of laser sources has however offered an alternative approach to studying the interaction of electromagnetic radiation with the surfaces of solids; in this context the photon source is coherent (with every wave being in-phase with each other), as opposed to being incoherent (a random relationship of one photon to each of the others emitted simultaneously). The main advantages of a laser in such studies are the enormous gains in beam intensity that can be achieved, and the ability to use controlled pulses of radiation varying in lifetime from ~ 1 ps upwards, however uv-emitting lasers are expensive relative to conventional gas-discharge (arc) systems and consequently this area of study has been relatively neglected.

2   THE INTERACTION OF ELECTROMAGNETIC RADIATION WITH SOLIDS

The interaction of near-u.v. photons with a typical photocatalytic material (e.g., finely divided titanium dioxide powder) is illustrated in Figure 3.1. The shape of the absorption edge reveals to some extent the shapes of the densities of electronic states in the valence band and in the adjacent conduction band and a diagrammatic representation of this situation is illustrated in Figure 3.2(a). The onset of the absorption edge (lowest energy) indicates the minimum energy required to effect the excitation of an electron from the highest occupied energy state of the valence band into the lowest available energy state of the conduction band. This transition may not necessarily involve the coupling of the electric vector of the photon directly with the wave functions of the two states involved—in cases where this occurs this process is termed a direct transition and results in a small change ($\Delta \mathbf{k} \approx 0$) in the wave vector of the electron involved following its excitation to the conduction band. More frequently, the magnitude of the wave vector ($\mathbf{k}'_{cb}$) of the lowest energy state in the conduction band does not correspond to that of the electron at the highest energy of the valence band ($\tilde{\mathbf{k}}_{vb}$), and as a consequence of the significant difference in these wave vectors ($\Delta \tilde{\mathbf{k}}$) requires that the excitation of the electron from the lower state (vb) to the highest state (cb) is accompanied by the participation, simultaneously, of an elastic wave within the lattice of the solid (a phonon) (illustrated in Figure 3.2(b)).

In such circumstances the transition probabilities of the two processes differ significantly in magnitude, a useful analogy being with two-body (direct) and three-body (indirect) collisions which occur between gaseous molecules. In the cases described, the 'bodies' are the photon and the electron in the direct transition or the photon, the phonon and the electron in the indirect transition.

At photon energies higher in magnitude than the minimum value required to effect transitions of the types described, electrons from the highest occupied states in the valence band can be promoted to higher energy states within the conduction band or into other unoccupied bands that may exist at even larger energies. In

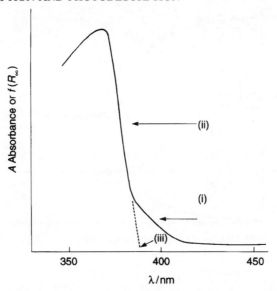

Figure 3.1. Diffuse reflectance spectrum of the fundamental absorption edge of titanium dioxide powder. Absorbance ($f(R_\infty)$) vs. wavelength: (i) indirect transition; (ii) direct transition; (iii) direct band gap; $E_g = Nhc/\lambda_g$.

general, the range of energies (ΔE) of the conduction band from its lowest value to the highest will be greater than the corresponding range of energies of the valence band; and moreover, while the onset of absorption may be due to an indirect process with a correspondingly smaller value of $-dA/d\lambda$, direct processes with larger $-dA/d\lambda$ values will be likely at higher photon energies. Additionally the absorption of higher-energy photons may raise electrons from lower-energy states in the valence band to the lower regions of the conduction band, but the accompanying hole state in the valence band will have a larger momentum than one created nearer to the top of the band.

In summary, therefore, the absorption of a photon at the fundamental absorption edge of the solid creates a hole state in the valence band and an occupied electronic

$$MX(s) + h\nu(\geqslant Eg) \rightarrow h^+_{v.b.} \text{ --- } e^-_{cb} \quad \text{exciton formation}$$
$$h^+_{v.b.} \text{ --- } e^-_{cb} \rightarrow h\nu' \text{ or } h\omega \quad \text{relaxation}$$

where $\nu' < \nu$ and $h\omega$ is a quantum of lattice vibrational energy.

Scheme 3.1

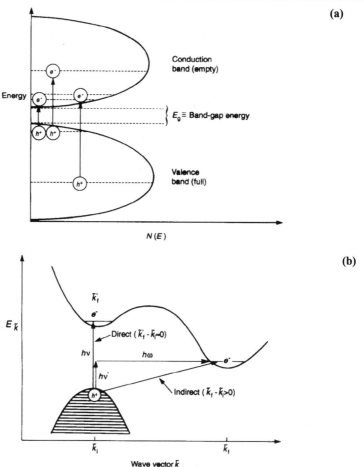

Figure 3.2. (a) Schematic energy band diagram showing hole–electron pair formation as a consequence of differing energies of incident radiation; (b) schematic energy band diagram (energy vs. wave vector **k**) illustrating the difference between direct and indirect transitions.

state in the conduction band which, when their difference of wave vectors is small, remain bound together by the mutual Coulombic attraction forces between them and constitute a hole–electron pair, or exciton (Figure 3.3(a)).

Hole–electron pairs exist only for short times, being effectively an excited state of the crystal lattice. The most common fate of hole–electron pairs is for them to recombine with the release of energy either as photons (luminescence) or, more commonly, as phonons to provide additional vibrational energy to the lattice centres of the crystal.

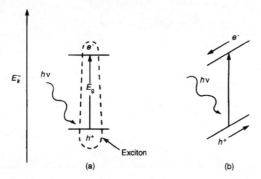

(a)   (b)

Figure 3.3. (a) No applied electric field; absorption of a photon (hv) to create an exciton, the electronic spectrum of which is similar in form to that of the spectrum of atomic hydrogen; (b) applied electric field of sufficient intensity to separate the excitonic charges and to accelerate them ($\rightarrow$) in opposite directions in $\mathbf{k}$-space.

3 CHARGE CARRIER SEPARATION

In the presence of an applied electrical field, of sufficient intensity to provide energy to both the hole (in one direction in k space) and to the electron (in the opposite direction in k space to that of the hole movement) to overcome the Coulombic attraction forces between them, the exciton may be dissociated into a quasi-free hole in the valence band and a quasi-free electron in the conduction band which now have a much lower tendency to recombine because of the increased difference in wave vector, $\tilde{\Delta}\mathbf{k}$. (Figure 3.3(b))) The effect of the applied electrical field is to increase greatly the lifetimes of the charge carriers and to offer opportunities for them to participate in other types of interaction. Interactions may occur with the ionic states which constitute the crystal lattice itself, as, for example, in the case of the silver halides, in which the electrons reduce the monovalent silver cations to silver atoms, while the quasi-free holes (positive holes) oxidise the halide centres to produce free halogen atoms. These steps are summarised in Scheme 3.2 which effectively describes the creation of the latent silver image in the photographic process.

In the current language of photocatalysis, solids which exhibit these internal forms of oxidation/reduction process are said to undergo 'photocorrosion' particularly when they occur in liquid suspension. When such processes occur with the solid *in vacuo*, the release of the gas molecules can be distinguished from photodesorption from surface species by the kinetics of their release, and by the extent to which they occur, photodesorption being effectively confined to less than one monolayer in extent, while the process previously described can, in principle, extend to well in excess of the monolayer limitation.

$$AgX(s) + h\nu \longrightarrow h^+_{v.b.} \longrightarrow e^-_{cb}$$

$$e^- + Ag^+_{lattice} \xrightarrow[\text{step}]{\text{reduction}} Ag^o \qquad \text{Silver atom formation}$$

$$h^+ + X^- (Cl^-, Br^- \text{ or } I^-) \xrightarrow[\text{step}]{\text{oxidation}} X(ads) \qquad \text{Halogen atom formation}$$

$$X(ads) + X(ads) \longrightarrow X_2(ads)$$

$$X_2(ads) \longrightarrow X_{2(g)} \qquad \text{Molecular desorption}$$

$$Ag^o + Ag^o \longrightarrow (Ag_2)^o \qquad \left\{ \begin{array}{l} \text{Nucleation} \\ \text{of} \\ \text{silver cluster} \end{array} \right.$$

$$(Ag_2)^o + Ag^o \dashrightarrow (Ag_3)^o$$

Scheme 3.2

4   CHARGE CARRIER RECOMBINATION

Hole–electron recombination in the solid can result in luminescent emission (Figure 3.4(a)) the quenching of which can be effected by the presence of a gaseous atmosphere, this topic is discussed authoritatively by Anpo et al. [6, 7]. Alternatively, the recombination process can result in the creation of a vibrational wave (phonon) in the crystal lattice. These phonons may result in increased displacements of the lattice centres in a direction perpendicular to the direction of motion of the wave (transverse phonon) or can lead to enhanced displacements of the lattice centres along the direction of travel of the wave (longitudinal phonon) (Figure 3.4(b)). The emergence of these phonons at the surface can result in the bonding of surface atoms (or adatoms) being weakened sufficiently for them to be ejected from the surface with excess kinetic energy (Figure 3.4(c)). In the normal course of events the desorption of species from the surface would be expected to exhibit a Maxwell- -Boltzmann distribution- of kinetic energies and be subjected to the directional limitations imposed by the cosine law relationship on each element of surface area (Figure 3.5).

Excited states of the lattice created by recombination processes may not relax quickly enough to be Maxwellian in their behaviour with the result that the distribution of kinetic energies may be non-Maxwellian and the cosine relationship may be violated due to the directional constraints on phonon movement imposed by the crystal lattice. Such information can be deduced from pulsed-laser-time-of-flight mass spectrometry studies as have been illustrated by the work of Polanyi [8–10]; this topic is however one for which much further work can be anticipated.

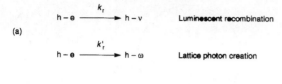

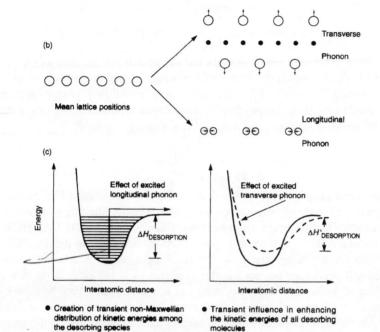

Figure 3.4. (a) Hole–electron recombination processes; (b) transverse and longitudinal phonons; (c) consequences of phonon interactions with surface species.

5 REVERSIBLE PHOTOADSORPTION AND PHOTODESORPTION: THE INFLUENCE OF ELECTROMAGNETIC RADIATION ON CHARGE-TRANSFER CHEMISORPTION EQUILIBRIA

An important factor, which arises as a consequence of the chemisorption of a reactive gas at the surface of the solid, particularly where the solid has semi-conducting character, is the effect which the chemisorption process exerts upon the position of the Fermi level (E_F) relative to the positions of the conduction band edge (E_C) and the valence band edge (E_V).

In situations in which the adsorption of an electrophile, such as dioxygen (O_2), takes place (Figure 3.6(a,b)) the Fermi level is depressed by an amount eV_s where

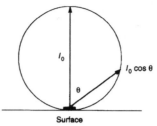

Figure 3.5. Randomised desorption of species from an element of area (dA) of the surface.

V_s is the surface potential created through the adsorption process and an electric field is created which restricts the access of electrons and encourages the access of holes to the surface. The result of this difference in accessibility of charged species provides a displacement of the position of the thermal equilibrium by desorbing molecules of adsorbed species according to the following equation:

$$h_{vb}^+ + O_2^- (\text{ads}) \rightarrow O_{2(\text{ads})}$$
$$O_2(\text{ads}) \rightarrow O_{2(g)}$$

A new equilibrium position is established as the Fermi level is raised by the desorption process until a steady state is re-established, whereby the rate of arrival of electrons and holes at the surface again becomes equal (Figure 3.6(c)). Conversely, if the transfer of electronic charge takes place from the adsorbate to the adsorbent, then the Fermi level will be raised and the resulting electric field in the space-charge layer will encourage the net transfer of electrons to the surface via downwards bending of the bands if hole–electron pairs are generated within the solid by the absorption of photons of appropriate energy. In such circumstances one can anticipate also the desorption of a neutral species as a consequence of the interaction of a conduction band electron with a positively charged adsorbed species. In each of the situations that have been described, discontinuation of the photon flux will result in the system relaxing towards the thermal equilibrium position produced before the system was irradiated. Under these circumstances a reversible change can be induced, i.e., adsorption under irradiation (photoadsorption) and desorption in the dark to re-establish the original position of thermal equilibrium or desorption under irradiation (photodesorption) and re-adsorption in the dark. An example of this behaviour is typified by zinc oxide in which both types of reversible behaviour have been observed as a result of different thermal pretreatments of the zinc oxide powdered specimen prior to the establishment of the initial thermally activated equilibrium condition [5]. Zinc oxide is an n-type oxide but the position of the Fermi level relative to the upper edge of the valence band (E_V) and the lower edge of the conduction band (E_C) is dependent upon the extent of the deviation of its composition from the (1 : 1) stoichiometric value (Figure 3.7(a,b)). Thermal conditioning of the oxide *in vacuo*, or in a reducing atmosphere such as in hydrogen gas, results in an excess of the metallic component (Zn^0) which raises the Fermi

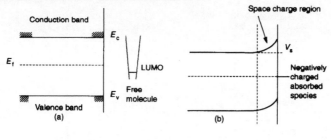

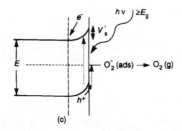

Figure 3.6. (a) Band structure of solid *before* its interaction with an electrophile; (b) band structure of solid *after* its interaction with an electrophile (thermal equilibrium); (c) influence of incident band-gap radiation on the adsorption equilibrium photodesorption.

level closer to the conduction band edge, i.e., the difference in energy $E_C - E_F$ is less for the reduced solid than it is for the oxidised form.

6 IRREVERSIBLE OXYGEN PHOTOADSORPTION AT TITANIUM DIOXIDE SURFACES

Photoadsorption and photodesorption effects are not confined exclusively to the category described, although these may be attributed to the intrinsic properties of a solid of defined composition and a gaseous (adsorbate) species. In some respects a much more significant source of photosorption arises from the existence of residual impurity species which originate from the preparative methods used to obtain the finely divided solids. Metal oxides are only prepared infrequently in a finely divided state by the combustion of a massive specimen of the parent metal (e.g., titanium) in pure oxygen followed by the comminution of the resulting oxide specimen by crushing or milling. Dispersed forms of oxides are more easily prepared by the formation of an insoluble hydrated oxide phase (precipitate) as a consequence of the hydrolysis of a reactive compound in water (titanium (IV) chloride, titanium (IV) sulphate, titanium (IV) alkoxide; titanium (III) chloride). The insoluble hydrated oxide phase is then carefully dried followed by calcining at

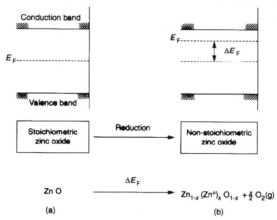

Figure 3.7. (a) Band structure diagram of stoichiometric ZnO; (b) band structure diagram of non-stoichiometric $Zn_{1+x}O$.

elevated temperature (e.g., 400–500°C) in a current of dry air or dioxygen to give a powder of large specific surface area. Even after such treatments, it is seldom the case that the surface of the resulting powder is entirely free from species such as adsorbed molecular water, hydroxyl species (Brønsted acids and bases), halide ions etc. and in restoring the specimen to ambient conditions re-adsorption of molecular water can occur fairly rapidly. In conditions of this type, photosorptive processes at the surface of oxides prepared by these methods can arise from the existence of additional types of charge-carrier trapping centres which offer alternative mechanisms by which photoadsorption in particular can occur. The photoadsorption of oxygen on the surface of powdered titanium dioxide at room temperature increases the surface coverage (θ) of oxygen by at most a few percent ($< 5\%$) of a monolayer [11]. The process requires the sustained flux of incident radiation of band-gap energy and does not undergo a spontaneous reversal when the photon flux is extinguished, although a proportion of the adsorbed oxygen can be removed by evacuation at ambient temperature. Increasing the temperature across the range, 0–50°C, provides a progressive increase in the net rate of photo-uptake of oxygen by the specimen; such characteristics imply that this form of photosorption is not merely a photo-induced elementary step but must consist of more than one elementary step, only one of which is photo-induced or, alternatively, at least one of which is thermally controlled and the execution of which imposes the limitation upon the net photo-uptake rate.

Different thermal outgassing pretreatments of powdered titanium dioxide at progressively higher temperatures, followed by reoxidation, bring about reductions in the capacity of the surface to photoadsorb oxygen under band-gap energy u.v. radiation. Studies of the thermal treatment processes by temperature programmed

$$TiO_2(s) + h\nu(\geqslant E_g) \xrightarrow{\hspace{2cm}} h^+ - e^-$$

$$h^+ - e^- \xrightarrow[\text{charge}]{\text{Space}} h^+_{vb} + e^-_{cb}$$

$$h + +OH^-_s \xrightarrow{\hspace{2cm}} \cdot OH_s$$

$$e^- + O_2(ads) \xrightarrow{\hspace{2cm}} O^-_2(ads)$$

$$\cdot OH_s + \cdot OH_s \xrightarrow{\hspace{2cm}} H_2O_2 \ (ads)$$

Scheme 3.3

desorption/mass spectrometry reveal the release of molecular water in three main regions of temperature ((i) 100–200°C, (ii) 250–325°C and (iii) > 400°C). The release of molecular water arises initially from strongly bonded molecular water (i), and from the interactions of weakly bound (ii) and more strongly bound (iii) Brønsted acid and base groups which, as a consequence of their interactions, release molecular water and create rearrangements to the remaining surface groups. Thermo-desorption studies of the hydrated surface of titanium dioxide *after* the photoadsorption of oxygen show that surface hydroxyl species, in particular, are active participants in the photosorption process and create a new low desorption energy state from which water and dioxygen are released together. Photoadsorption of oxygen on these surfaces leads to the progressive erosion of the water desorption peaks (iii) and (ii) respectively over a period of about 24 h. Conclusions to be drawn from these observations are that an extrinsically induced form of photo-adsorption is due to the participation of adsorbed hydroxyl species which result in the photoadsorption of molecular oxygen and the formation of adsorbed hydrogen peroxide. The mechanism for the photoadsorption process is summarised in Scheme 3.3.

The academic debate must revolve around whether it is strictly accurate to describe the process as photoadsorption when it results in the photo-oxidation of adsorbed water forming adsorbed hydrogen peroxide. The justification for its classification as photoadsorption relies exclusively upon the fact that the adsorbed coreagent in each of its distinctive forms ($H_2O(ads)$, $OH^-(ads)$) is strongly pre-adsorbed and that the products also remain adsorbed, albeit in weaker states of adsorption.

To conclude this discussion of oxygen photoadsorption on titanium dioxide surfaces, it must be emphasised that the kinetics of the photoadsorption process are dictated by the availability of hole traps (OH^- species) and as the surface is progressively dehydroxylated the rate of oxygen photoadsorption is reduced.

7 OXYGEN PHOTOADSORPTION AND ITS ROLE IN PHOTOCATALYSIS

An augmentation to the rate of oxygen photo-uptake by a TiO_2 surface can be created if the surface is pre-exposed to the vapour of a strongly adsorbing species, such as an aliphatic alcohol (ROH). Alcohols adsorb very strongly up to monolayer coverage on TiO_2 and the adsorbed species (e.g., RO^-(ads)) can themselves act as very efficient hole traps. In these circumstances the rate of uptake of oxygen by the surface under band-gap energy illumination is greatly enhanced and attains a pseudo-zero order character. In this case the surface becomes covered progressively with the oxidised form of the alcohol (ethanal when ethanol is used; propanone when propan-2-ol is used).

Accordingly, a study of strongly preadsorbed organic species in the presence only of a pure gaseous species such as $O_2(g)$ is strictly a precursor to sustained photocatalytic oxidation where, with the availability also of excess vapour or liquid phase reagent, a photocatalytic oxidation can occur with the progressive con-sumptions of both $O_2(g)$ (or O_2 (dissolved)) and the organic reagent e.g., propan-2-ol. In many examples, the kinetics of the photocatalytic reaction are controlled uniquely by the photo-electronic events induced in the solid in the copresence of the reacting partners.

8 OXYGEN PHOTO-EXCHANGE DURING PHOTOSORPTION PROCESSES

In earlier sections of this article, it is indicated that photodesorption or photo-release of species from the surface of a solid at thermal equilibrium with the gaseous phase can originate from either adsorbed oxygen species or from the oxygen species which comprise the crystal lattice of the solid. Clearly by labelling, isotopically, the oxygen gas differently from the lattice oxygen, it is possible on the one hand to study the scrambling of the gaseous isotopic mixture (e.g., $^{18}O_2(g)/^{16}O_2(g)$) or the contribution made to the hole-trapping by the oxygen anions of the oxide lattice ($^{18}O_2(g)/^{16}O^{2-}_{lattice}$). Such studies have been performed to good effect by Pichat and his co-workers [12].

$$^{18}O_2(g) + {}^{16}O_2(g) \xrightarrow[TiO_2]{hv} 2\,^{18}O^{16}O(g) \qquad \text{Molecular isotopic exchange}$$

$$^{18}O_2(g) + {}^{16}O^{2-}_{(lattice)} \xrightarrow[vTiO_2]{h} {}^{18}O^{16}O(g) + {}^{18}O^{2-}_{(lattice)} \qquad \text{Exchange with lattice anions}$$

However, oxygen exchange can emerge also in photosorption studies at hydrated titanium dioxide surfaces in which adsorbed hydrogen peroxide is a product. Adsorbed hydrogen peroxide is itself unstable under u.v. irradiation with band-gap photons on titanium dioxide surfaces with dioxygen being released as a con-sequence. Clearly in this situation isotopic scrambling of gaseous dioxygen with the oxygen in water should also contribute to the overall process (Scheme 3.4). Such definitive isotopic experiments have not yet been performed as far as this author is aware.

ISOTOPIC EXCHANGE OF OXYGEN WITH SURFACE HYDROXYL SPECIES

$$^{18}O_2(g) \rightleftharpoons\ ^{18}O_2(ads)$$

$$^{18}O_2(ads) + e^- \quad (\text{from solid}) \longrightarrow\ ^{18}O_2^-(ads)$$

$$^{18}O_2^-(ads) +\ ^{16}OH \xrightarrow[TiO_2]{hv}\ ^{18}O^-(ads) +\ ^{18}O^{16}OH(ads)$$

$$H^{16}O_2 \cdot (ads) + H^{16}O^{18}O \cdot (ads) \longrightarrow H^{16}O^{16}OH(ads) +\ ^{16}O^{18}O(ads)$$

or

$$\searrow\ H^{16}O^{18}OH(ads) +\ ^{16}O_2(ads)$$

$$^{16}O_2(ads) \longrightarrow\ ^{16}O_2(g)$$

Scheme 3.4

9 SOME EXPERIMENTAL APPROACHES TO THE STUDY OF PHOTOADSORPTION AND PHOTODESORPTION

1. *Manometric studies*. The use of a high-vacuum system or ultra-high-vacuum system combined with a sensitive manometric device will enable overall changes of pressure in a system of constant volume to be determined with ease.

2. *Mass-spectrometric studies*. Any high-vacuum manometric measurements should not now be made in the absence of mass-spectrometric monitoring of the composition of the gas phase throughout the course of the experiment. In this manner, compositional changes can be interpreted in a more meaningful manner than by merely using manometric information, since it will reveal the release of species from the surface (e.g., H_2O) as a result of the irradiation process itself and, as described in Section 8, will permit mechanistic investigations which are not possible otherwise. A more recent development occurs through the use of time-of-flight mass spectrometry, whereby not only the mass

of the species can be measured but the kinetic energy also, enabling checks to be made upon whether species leaving the surface as a consequence of the absorption of a photon possess non-Maxwellian behaviour.

3. *Thermo-desorption studies.* Temperature-programmed desorption (tpd) in conjunction with the two previously mentioned techniques can facilitate an understanding of the changes that are brought about by irradiating a surface either without the presence of a gaseous phase (i.e., *in vacuo*) or in the presence of a gaseous phase as, for example, in the study of the photoadsorption of dioxygen on titanium dioxide surfaces. The important information is revealed by studying, firstly, the thermally equilibrated surface (by tpd) and then following this by studying an identical surface, having subjected it to the irradiation process.

4. *Surface spectroscopies and their impact upon the mechanistic aspects of photoadsorption and photodesorption.* Non-destructive spectroscopic methods can be adapted to examine the nature of adsorbed species. In general quanta of vibrational energy or of rotational energy are extremely small ($hv <$ 0.01–0.1 eV mol^{-1}) and transitions of this type which are present in surface molecular species can be probed at appreciable surface coverages ($\theta \geqslant 0.1$) by using Fourier transform infra-red spectroscopy (FTIR) of Fourier transform Raman spectroscopy (FTRS). Electron energy loss spectroscopy (EELS) or X-ray photoelectron spectroscopy (XPS) also permit investigations of surface species but they expose the surface to relatively high-energy quanta which can themselves induce changes to the species present, although time constraints may limit the extent of changes that can be induced relative to the time required for the measurement itself.

One of the most successful forms of spectroscopic study has been the application of electron spin resonance to the identification of free radical species (unpaired electrons) e.g.,

$$\cdot OH, \quad \cdot O_2H, \quad \cdot O_2^-, \quad \cdot O^-, \quad \cdot O_3^-.$$

The technique is well established and has been used with considerable success by Soria (in Madrid) [13], Vedrine (in Lyon) [14] and by Kazanski (in Russia) [15, 16].

One of the most under-used methods has been the application of u.v.-visible diffuse reflectance spectroscopy to study the fundamental optical absorption edge of the solid and the modification of it by the adsorption of gases. This technique has been developed very satisfactorily by Stone in the UK and by Zecchina and colleagues in Italy [17, 18]. It is a powerful tool, enabling surface electronic states to be identified, and to follow their modification as a consequence of being subjected to irradiation by photons of band-gap energy.

5. *Photoelectronic studies of solids and surface species.* One of the most difficult aspects of studies of photoadsorption and photodesorption relates to following the consequences of the absorption of photons to produce an electrically excited state of the solid, or to produce subsequently highly reactive surface radical species. In the former situation, there is a transient increase in the number of charge carriers which will be rapidly restored to the initial condition via charge-carrier recombination if the source of photons is extinguished. Such processes may be studied by following the change in Q-factor of a tuned cavity containing the specimen which is being subjected simultaneously to continuous irradiation with microwaves (GHz), and to short pulses (μs) of photons necessary to create the excited state. A modified esr spectrometer system can be used for this purpose and such experiments can assist in the understanding, firstly, of the kinetics of the recombination processes and secondly, of the role of impurity centres in relation to charge-carrier recombination [19–21].

In normal conditions (e.g., at room temperature) surface radical species will possess very short relaxation times which will lead to excessively large line-widths. In these circumstances information on the identities of the radicals can frequently be achieved by reducing the temperature at which the measurements are made to below *ca.* 100 K. Under these conditions the relaxation times become much longer and the spectral features are more easily discerned. Whereas, even at 77 K, hole–electron recombination lifetimes can be of the order of 1×10^{-9} s to 1×10^{-6} s, under similar conditions of temperature, surface radical species can be studied for several minutes.

10 CONCLUDING REMARKS

Photoadsorption and photodesorption frequently are merely small parts of the richly intellectual subject of photocatalysis, the development of which commenced in 1910 (Prudnikov) and proceeded slowly until *ca.* 1966–70. Since that period, stimulated mainly by the photoelectrochemical experiments of Fujishima and Honda in Japan, a resurgence of interest has been sustained through, for example, aspirations to photocatalyse the electrolysis of water to produce cheap supplies of hydrogen as a fuel; and through the use of photocatalysis to remove low concentration levels of organic pollutants from water. Such challenging tasks require improved understandings of the means of converting u.v. photons into forms of energy which can promote chemical processes. The foregoing pages have sought to illustrate ways in which this task has been approached for the subjects of photoadsorption and photodesorption. It is clear to the author that real progress can be achieved by viewing the problem only through the simultaneous use of at least two distinctive experimental techniques, and preferably more, in order to provide a wide perspective to the acquisition of self-consistent explanations to all of the observable information on the topic.

REFERENCES

[1]   Brunauer S., Deming L., Deming W. S., and Teller E., Classification of the Shapes of Adsorption Isotherms, *J.A.C.S.* **62**, 1723 (1940).

[2]   Brunauer S., Emmett, P. H., and Teller E., The BET Theory of Physical Adsorption, *J.A.C.S.* **60**, 309 (1938).

[3]   Gregg S. J. and Sing K. S. W., *Adsorption, Surface Area and Porosity.* Second Edition. Academic Press, New York (1982).

[4]   Bickley R. I., Photoadsorption and Photodesorption at the Gas–Solid Interface. Part I. Fundamental Concepts, *Photocatalysis and Environment.* Edited by M. Schiavello. NATO ASI, Series C Vol. 237, pp. 223- -232, Kluwer Academic Press, Dordrecht (1988).

[5]   Bickley R. I., Photoadsorption and Photodesorption at the Gas–Solid Interface. Part II. Photoelectronic Effects relating to Photochromic changes and to Photosorption, *Photocatalysis and Environment.* Edited by M. Schiavello. NATO ASI, Series C Vol. 237, pp. 233–239, Kluwer Academic Press, Dordrecht (1988).

[6]   Anpo M., *Surface Photochemistry*, Wiley, Chichester (1996).

[7]   Anpo M., Chiba K., Tomonari M., Coluccia S., Che M., and Fox M.A., Photocatalysis on Native and Pt-Loaded TiO_2 and ZnO Catalysts—origin of different reactivities on wet and dry metal oxides, *Bull. Chem. Soc. Japan* **64** (2), 543–551 (1991).

[8]   Bourdon E. B. D., Cowin J. P., Harrison I., Polanyi J. C., Segner J., Stanners C. D., and Young P. A., Photodissociation, Photoreaction and Photodesorption of Adsorbed Species. Part I. Methyl Bromide on a Lithium Fluoride (001) Surface, *J. Phys. Chem.* **88**, 6100 (1984).

[9]   Bourdon E. B. D., Das P., Harrison I., Polanyi J. C., Segner J., Stanners C. D., Williams R. J., and Young P. A., Photodissociation, Photoreaction and Photodesorption of Adsorbed Species. Part II. Methyl Bromide and Hydrogen Sulphide on a Lithium Fluoride (001) Surface, *Faraday Disc. Chem. Soc.* **82**, 343–358 (1980).

[10] Dixon-Warren S. J., Jensen E. T., Polanyi J. C., Xu G.-C., Yang S. H., and Zeng H. C., Photochemistry of Adsorbed Molecules. Part 10. Charge Transfer from Ag or K substrates to Halide Adsorbates, *Faraday Disc. Chem. Soc.* **91**, 451–463 (1991).

[11] Bickley R. I., Day R., Jayanty R. K. M., Navio J. A., Payne C., and Vishwanathan V., The Role of Photo-electronic Processes in the Formation of Active Oxygen during the Photo-fixation of Dinitrogen on Titanium Dioxide Surfaces, *Proceedings of the Ninth International Congress on Catalysis.* Edited by M. Philips and M. Ternan. Vol. 4, pp. 1505–1515. Chemical Institute of Canada (1988).

[12] Pichat P., Powder Photocatalysts; Characterisation by Isotopic Exchange and Photoconductivity, *Photocatalysis and Environment.* Edited by M. Schiavello. NATO ASI, Series C Vol. 237, pp. 399–424, Kluwer Academic Press, Dordrecht (1988).

[13] Soria J. and Gonzalez-Elipé A. R., E.S.R. Study of the Radicals formed by u.v.-irradiation of TiO_2 in the presence of SO_2 and O_2, *J. Chem. Soc. Faraday Transactions (I)* **82**, 739–745 (1986).

[14] Meriaudeau P. and Vedrine J., Paramagnetic Oxygen Species on Titanium Dioxide Surfaces, *J. Chem. Soc. Faraday Transactions (I)* **72**, 472 (1976).

[15] Shvets V. A. and Kazanski V. B., Paramagnetic Oxygen Species on Irradiated Magnesium Oxide Surfaces, *J. Catalysis* **25**, 123 (1972).

[16] Nikisha V. V., Selimov B. N., and Kazanski V. B., Paramagnetic Oxygen Species on Titanium Dioxide Surfaces, *Kinetics and Catalysis* **15**, 599 (1974).

[17] Stone F. S., Garrone E., and Zecchina A., Surface Properties of Alkaline Earth Oxides as studied by Diffuse Reflectance Spectroscopy, *Mater. Chem. Phys.* **13** (3/4), 331–346 (1985).

[18] Zecchina A. and Stone F. S., Diffuse Reflectance Studies of Alkaline Earth Oxide Surfaces, *J. Chem. Soc. Faraday Transactions* **72**, 2364 (1976).

[19] Kunst M. and Beck G., Microwave Conductivity Measurements, *J. Appl. Phys.* **60** (10), 3558–3566 (1986).

[20] Kunst M. and Beck G., Microwave Conductivity Measurements, *J. Appl. Phys.* **63** (4), 1093–1098 (1988).

[21] Kunst M. and Schindler K-M., Charge Carrier Kinetics in Zinc Oxide Powders, *Z. Naturf.* **43a**, 189–192 (1988).

4 Thermodynamics and Kinetics for Heterogeneous Photocatalytic Processes

L. PALMISANO and A. SCLAFANI

Dipartimento di Ingegneria Chimica dei Processi e dei Materiali, Università di Palermo, Viale delle Scienze, 90128 Palermo, Italy

1 INTRODUCTION

Photocatalysis has gained more and more attention by scientists all over the world and many books [1–5] reporting studies on several processes successfully carried out by using polycrystalline semiconductors as catalysts, both in gas–solid and in liquid–solid regimes, have been published in the last years. The most widely used semiconductor is polycrystalline TiO_2 in the anatase or rutile phase, due to its (photo)-stability, to the not high band-gap value (≈ 3.0 and ≈ 3.2 eV for rutile and anatase, respectively) and to the fact that it doesn't present any toxicity.

Heterogeneous Photocatalysis, Edited by M. Schiavello
© 1997 John Wiley & Sons Ltd.

The knowledge of the main thermodynamic and kinetic factors influencing the photoreactivity is essential in order to make predictions on the feasibility of photoprocesses and to explain why some solids are active and others are not. Moreover the same solids can be active in a particular system for a particular photoreaction and not active in different experimental conditions or for different reactions. Therefore, thermodynamic and kinetic factors governing the photocatalytic processes should be carefully considered in order to choose the best experimental conditions for their occurrence.

The first part of the present chapter will deal with the main thermodynamic constraints governing the feasibility of photocatalytic processes, and, in particular, the consequences of the illumination of polycrystalline semiconductors in contact with a fluid and with redox couples will be considered.

Some selected photo-reactions widely studied in recent years, as for instance dinitrogen photoreduction to ammonia [6–9], carbon dioxide photoreduction to organic compounds [10–17], water photosplitting [18–20] and the partial or complete photo-oxidation of organic and inorganic [3,5,21–27] species in aqueous medium will be discussed. The above processes can be divided into two classes: the first class includes spontaneous reactions ($\Delta G < 0$), the second one non-spontaneous reactions ($\Delta G > 0$).

Moreover, the influence on the photoreactivity of the preparation method and the crystalline phase of the catalyst along with some kinetic factors will be discussed; i.e. (i) the recombination rate of the photoproduced pairs; (ii) the adsorption–desorption of reagent species on the catalyst surface; (iii) the charge transfer from the surface of the solid to the reagents. The above factors can be influenced for instance by the presence of some group VIII metals or Ag deposited on the surface of the catalyst particle and by the presence of some inorganic species such as Ag^+, Fe^{2+}, Fe^{3+}, Cr^{3+}, Cu^{2+} and H_2O_2, used as additives in the reaction medium or as dopants for the photo-catalyst, in particular TiO_2. The reaction rates can be positively or negatively influenced, depending on the type of reaction (photo-reduction or photo-oxidation), on the regime (gas–solid or liquid–solid) and on the location of the additives (in the solid photocatalyst or in the bulk of the reaction medium). It should be taken into account by the reader that both oxidation and reduction reactions occur on the surface of the catalyst during a photocatalytic process. Consequently, the terms photo-reduction and photo-oxidation have to be referred to the chemical transformation of the main reagent species.

2 DEFINITION OF HETEROGENEOUS
   PHOTOCATALYTIC PROCESSES

Heterogeneous photocatalytic processes can be defined as catalytic processes during which one or more reaction steps occur by means of electron–hole pairs photogenerated on the surface of semiconductor materials illuminated by light of suitable energy.

The above general and wide definition implies that some steps of a photo-catalytic process are redox reactions involving the photogenerated electron–hole (e^-–h^+) pairs.

Some authors distinguish between 'photocatalytic' and 'photosynthetic' processes, according to whether the ΔG values of the global reactions are smaller or greater than zero, respectively [28].

According to the above distinction, 'photocatalytic processes' are spontaneous reactions ($\Delta G < 0$) occurring with higher rate when particular semiconductor solids illuminated by light of suitable energy are present. Different reaction pathways with a lower activation energy, indeed, can be created under illumination.

We cite, for instance, the oxidation of propan-2-ol in the presence of O_2:

$$CH_3\underset{\underset{OH}{|}}{CH}CH_3 + \tfrac{1}{2}O_2 \quad \xrightarrow[TiO_2]{h\nu} \quad CH_3COCH_3 + H_2O \tag{4.1}$$

The 'photosynthetic processes', instead, are not spontaneous reactions ($\Delta G > 0$) occurring only in the presence of particular semiconductor solids illuminated by light of suitable energy. A fraction of the photon energy absorbed by the system, in this last case, is converted into chemical energy.

The formation of propanone from propan-2-ol dehydrogenation and the production of an amino acid, i.e. glycine by methane and ammonia in the presence of water, are examples of this kind of process:

$$CH_3\underset{\underset{OH}{|}}{CH}CH_3 \quad \xrightarrow[Pt/TiO_2]{h\nu} \quad CH_3COCH_3 + H_2 \tag{4.2}$$

$$2CH_4 + NH_3 + 2H_2O \quad \xrightarrow[Pt/TiO_2]{h\nu} \quad H_2NCH_2COOH + 5H_2 \tag{4.3}$$

The distinction above reported does not highlight the role played by the semiconductor (generally an oxide) in breaking and creating chemical bonds.

The occurrence of the photo-splitting of water according to the following reaction:

$$H_2O \quad \xrightarrow{h\nu} \quad H_2 + \tfrac{1}{2}O_2 \tag{4.4}$$

needs photons with $\lambda < 186\,nm$ ($E_1 = h\nu_1$) while in the presence of particular semiconductor oxides, e.g. TiO_2, the required energy is far lower, i.e. only equal to the band-gap energy of the semiconductor, in this case $3.2\,eV$ ($E_2 = h\nu_2$) corresponding to $\lambda \cong 390\,nm$.

The difference between the two values ($E_1 - E_2$) is due to different reaction mechanisms and gives indication about the catalytic role of TiO_2 in reaction (4.4).

When a semiconductor is used with a band-gap value smaller than that of TiO_2 and with conduction and valence bands thermodynamically compatible with the reaction of photodecomposition of water, a reaction rate higher or lower compared with that observed for TiO_2 could be observed. In the case of lower rate the semiconductor would be a photocatalyst less efficient than TiO_2, independently from the energy needed for its activation.

Moreover, by considering TiO_2 and $TiO_2 \cdot Pt$, they show the same band gap and the photodecomposition of water is thermodynamically possible in the presence of both of them. The reaction rate, though, is higher in the presence of $TiO_2 \cdot Pt$ as the presence of Pt, in this case, is responsible for the enhancement of the catalytic activity.

We are inclined to use the more general definition reported in the beginning of this section where the term 'photocatalytic' has been used only for spontaneous reactions. Catalytic processes are, indeed, substantially independent of thermodynamic parameters and *photocatalytic reactions are those for which the rate increases as a consequence of an optical excitation of a solid that remains chemically unchanged.*

It is, however, convenient to define the endergonic reactions ($\Delta G > 0$) as 'photosyntheses' or 'catalytic photoassisted syntheses', CPS, and the exergonic reactions ($\Delta G < 0$) as 'catalytic photoassisted reactions', CPR. It is worth noting, indeed, that some natural reactions photoassisted by solar light, such as the chlorophyllic photosynthesis, have been historically called 'photosyntheses'.

3 THERMODYNAMICS OF HETEROGENEOUS PHOTOCATALYTIC PROCESSES

When an n-semiconductor is illuminated by light of suitable energy, electrons are promoted from the valence band to the conduction band and positive holes are created in the valence band. The occurrence of the above charge separation is one of the first essential steps of a photocatalytic reaction.

From a physical point of view, light absorption, therefore, induces the formation of electron–hole pairs. The photoproduced pairs can subsequently evolve in different ways: (i) they can recombine with emission of thermal energy and/or luminescence or (ii) react with electron acceptor or donor species giving rise to reduction and oxidation processes, respectively.

The potential energy of the photogenerated electrons and holes can be given up to a redox couple with suitable redox potential, i.e. less positive than that of the valence band of the semiconductor (holes oxidise the reduced species of the couple) and more positive than that of the conduction band of the semiconductor (electrons reduce the oxidised species of the couple), as can be seen in Figure 4.1. In this figure the position of the band edges of some semiconductors are reported by using as reference the normal hydrogen electrode (NHE) scale.

More than one couple, of course, can be involved in a photocatalytic process, and electrons and holes can reduce and oxidise, respectively, species belonging to different redox couples contemporaneously present in the reacting system.

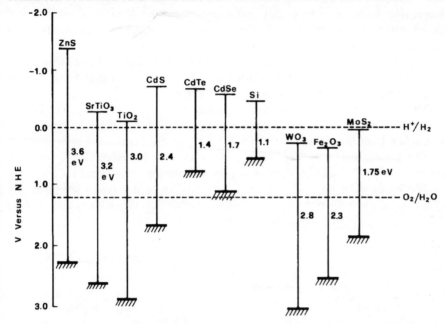

Figure 4.1.   Positions of the band edges for some semiconductors in contact with aqueous electrolyte.

The above phenomena, by means of which the semiconductor converts light energy in chemical energy, are of fundamental importance in photocatalysis.

A prevalence, of course, of the occurrence of redox processes on the recombination of pairs enhances the photocatalytic activity of the semiconductor.

When the same redox couple is considered, different reaction pathways could be observed in the presence of two different semiconductors with similar band gaps, depending on the position of their conduction and valence bands.

Hence, for instance, photooxidation of water in the presence of MoS_2 (band gap 1.75 eV) is possible from a thermodynamic point of view, but the same reaction is not possible in the presence of CdSe (band gap 1.70 eV) due to the position of the valence band of this solid. The photo-reduction of water, instead, could occur more easily in the presence of CdSe (see Figure 4.1), due to the position of its conduction band.

4   SEMICONDUCTOR IN THE PRESENCE OF A FLUID

When a semiconductor and a liquid solution are in contact (solid–liquid junction), a charge transfer occurs until an electrostatic equilibrium is achieved, i.e. their Fermi levels (E_f) reach the same energy and a depletion layer of major carriers is produced within the semiconductor, with a subsequent bending of the bands at the semiconductor–electrolyte interface.

If the Fermi level (E_f^o) of the semiconductor is higher than E_{redox} of the solution, electrons are transferred to the solution and the semiconductor and the solution acquire positive and negative charges, respectively.

The excess of charge in the semiconductor is not confined on the surface as occurs in a metal, but is distributed in a region called the 'space charge region' expanding from the surface to the bulk of the semiconductor.

The electric field resulting from the formation of the space charge region is responsible for the band bending of E_{cb} and E_{vb}, as shown in Figures 4.2 and 4.3.

The bands bend up when the semiconductors (n- and p-types) are positively charged, i.e. when E_f^o of the semiconductor is higher than the redox potential of the

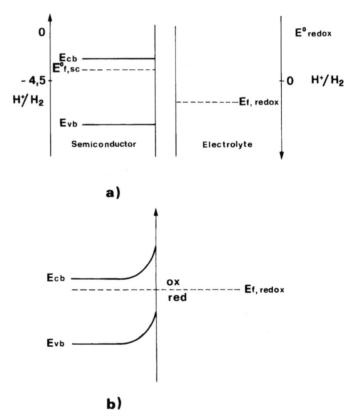

Figure 4.2. Energy diagram for the case $E^o_{f,sc} > E_{f,redox}$: (a) position before semiconductor–solution contact; (b) position under equilibrium condition, after semiconductor–solution contact.

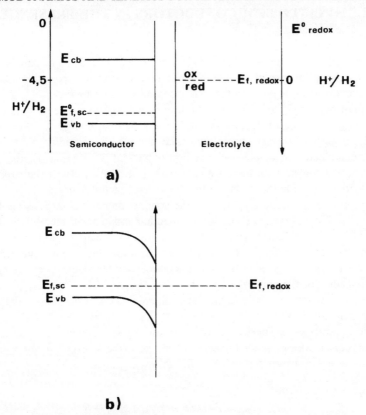

Figure 4.3. Energy diagram for the case $E^o_{f,sc} < E_{f,redox}$: (a) position before semiconductor–solution contact; (b) position under equilibrium condition, after semiconductor–solution contact.

couple(s) present in the solution (see Figure 4.2). For negatively charged semiconductors ($E^o_f < E_{redox}$) the situation is opposite (see Figure 4.3).

For positively charged n-type semiconductors, transfer of electrons occurs from the surface to the bulk of the particles and positive holes are confined on the surface (depletion layer), while for the negatively charged ones, electrons accumulate on the surface (accumulation layer).

The opposite situation arises for p-type semiconductors.

The above considerations can be extended to other cases, in particular to semiconductors in contact with gaseous phases. In this last case, the value of the potential difference depends on the extent of interaction between the two phases, i.e. on the extent of chemiadsorption and on the pressure of the gas.

5   ILLUMINATED SEMICONDUCTORS IN THE PRESENCE OF A FLUID

The illumination of a semiconductor with light of suitable energy, namely higher than its band-gap energy, gives rise to promotion of electrons from the valence to the conduction band, leaving in the valence band positive holes.

The photon energy stored by the photoproduced electron–hole pairs is equivalent to $E_{band\ gap}$, being the excess transformed in thermal energy.

In the presence of surface charge layers, i.e. a space-charge region, as in the case of a semiconductor in contact with a liquid phase (see Section 4), the photo-produced pairs can be separated. Consequently, the probability of their recombination decreases, while their availability for redox reactions increases.

The promotion of electric charges in the conduction band is responsible for the modification of the Fermi levels of the semiconductor and of the electrolyte (E_f^o and E_f, respectively). The equilibrium can be restored by means of a charge transfer with redox species present in the solution. In particular the photo-produced holes can oxidise the reduced species of the redox couple and the electrons can reduce the oxidised species of the couple.

The system works similarly to an electrochemical cell, continuously fed by the impinging photon flux.

If one assimilates the behaviour of an irradiated semiconductor to an electro-chemical cell working at steady-state conditions, the energy of the absorbed photons can be expressed according to the following equation:

$$\Delta E_{band\text{-}gap} = \Delta G/ZF + \Psi_{cat} + \Psi_{an} + Ir \tag{4.5}$$

where $\Delta E_{band\text{-}gap}$ represents the band gap of the semiconductor, ΔG is the free energy difference of the redox process occurring in the system, Z a positive integer number equal to the number of elementary charges involved in the redox process, F the Faraday constant, Ψ_{cat} and Ψ_{an} the total overvoltages (dissipative parameters) for the reductive and oxidative processes and the term Ir the ohmic drop in the electrolyte solution.

When the redox process is endergonic, as for instance the water photosplitting or the photosynthesis of organic compounds sketched in equation (4.6):

$$CO_2 + H_2O \xrightarrow[\text{semiconductor}]{h\nu} \text{organic compounds} \tag{4.6}$$

a portion of the photon energy is converted and stored as chemical energy. On the other hand, exergonic processes completely dissipate the available energy.

6 KINETICS OF HETEROGENEOUS PHOTOCATALYTIC PROCESSES

The rate of redox reactions occurring in a classical electrochemical system depends on the quantity of current circulating in that system for unit time, namely the current intensity, I. Consequently, by indicating as ΔV and Σr the voltage and the global 'reaction resistances' (i.e. the reaction hindrances), respectively, the rate of the process can be obtained from the value of $I(\Delta V/\Sigma r = I)$.

In a photocatalytic system the rate depends on the global 'reaction resistances' but also on the concentration of photoproduced electron–hole pairs. The concentration of these latter depends on the intensity of the radiation of suitable energy impinging on the system and on their recombination rate.

When the maximum concentration of the pairs has been achieved (steady state), the rate depends on several factors, such as electronic, chemical and morphological properties of the semiconductor, presence of additives in the reacting system, donor–acceptor and acid–base properties of the solution and of the solid, temperature and pressure.

One can follow the fate of the photoproduced electron–hole pairs by means of a simple sketch (Figure 4.4) in which the illuminated particle of a semiconductor, namely TiO_2, has been assimilated to a very simple electric circuit. It can be noticed that the availability of the electric charges in branch (b) is higher as the global resistance of branch (a) becomes more significant and/or the global resistance of branch (b) becomes less significant. This would mean a better utilisation of the photoproduced pairs with an increase of the quantum yield.

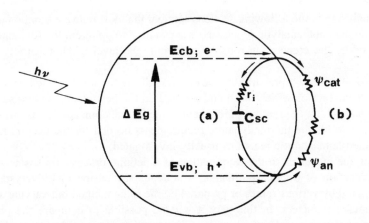

Figure 4.4. Illuminated TiO_2 particle assimilated to electric circuits. Circuit (a): capacities and resistances inside TiO_2 particle; circuit (b): resistances of the redox reactions and of the electrolyte. Symbols: Ψ_{cat} = cathodic overvoltage; Ψ_{an} = anodic overvoltage; r_i = internal resistance; r = electrolyte resistance; C_{sc} = semiconductor capacity; E_{cb} = conduction band energy; E_{vb} = valence band energy; ΔE_g = band-gap.

The values of resistance sketched as r in branch (b) of Figure 4.4 are not usually significant both for liquid–solid and gas–solid interfaces, at least when the semiconductors are in the polycrystalline form.

On the other hand, the values of Ψ_{cat} and Ψ_{an}, together with the surface physicochemical properties of the powders, strongly influence the 'catalytic' behaviour of the semiconductor.

The adsorption of reagent(s), the charge transfer from and to the reagent(s) and the desorption of the product(s) of the photoreaction are essential kinetic steps and their role should be evaluated every time as in catalysis.

These terms could be conventionally indicated as Ψ_{ads} (overvoltage due to adsorption), Ψ_{ct} (overvoltage due to charge transfer) and Ψ_{des} (overvoltage due to desorption).

The modification of some of the above steps can strongly influence the kinetics of photocatalytic processes and in the following sections some actions aimed to improve the photocatalytic activity of TiO_2 will be presented.

In principle, however, the actions to be undertaken in order to improve the photocatalytic activity of TiO_2 are not much different from those needed for other semiconductors.

7 INFLUENCE OF PREPARATION METHOD AND CRYSTALLINE PHASE OF CATALYST ON PHOTOACTIVITY

Before analysing in the following sections some of the most relevant kinetic factors influencing the photocatalytic processes, it is essential to point out the fundamental role played by the preparation route to determine the level of photoactivity of a catalyst.

We'll deal with two preparations of polycrystalline titanium dioxide as a case study. Indeed, TiO_2 is widely used and considered the best choice among several other oxides both for the reasons pointed out in the Introduction and for its low cost. Moreover it fulfils the thermodynamic requirements needed for the occurrence of most of the photocatalytic reactions usually investigated.

We confine ourselves to the examination of a definite process, for instance the photodegradation of organic pollutants in aqueous suspension of polycrystalline TiO_2 prepared by various routes or prepared by the same method but varying some experimental parameters. In this way it will be possible to compare the photoactivity of different powders and explain why catalysts, apparently identical by a chemical point of view, show different photoactivities.

When TiO_2 samples are prepared, for instance, starting from $TiCl_3$ or $TiCl_4$ by precipitating the precursor solids in aqueous medium by using NH_3 or $NaOH$, respectively, very different photocatalytic behaviours are observed [29] towards a probe reaction, namely phenol photodegradation in aqueous medium, well known in literature both from the mechanistic and kinetic point of view [21,23,25].

The samples deriving from $TiCl_3$ are obtained mainly in the anatase phase after a thermal treatment up to 873–923 K for 3–192 hours and all of them are photoactive, although in different degrees. The rutile phase can be obtained after more prolonged thermal treatments or by raising the temperature and the appearance of this phase is beneficial for the photoreactivity only if in mixture with the anatase phase. Indeed, when only rutile phase is present and it has been obtained at high temperature, namely after heating in air the precursor solid to 1073 K for 24 h, is the photoreactivity negligible.

The samples deriving from $TiCl_4$ show, instead, a significant amount of rutile phase after thermal treatments in air of their precursor solids at only 673–823 K for 3–24 hours. By raising the temperature (up to 873 K) and/or the duration of the thermal treatment, rutile phase becomes the unique phase or the dominant one. But, more significantly, apart from the structural differences due to the different preparations, the anatase samples obtained from $TiCl_4$ are much more photoactive than those obtained from $TiCl_3$.

Moreover, the samples obtained using $TiCl_4$ as starting reagent and containing mainly or exclusively rutile phase are the most photoactive when the thermal treatment does not exceed 973 K (3 hours). By raising the temperature and/or the duration of the thermal treatment, negligibly photoactive rutile is obtained, as for the samples prepared by using $TiCl_3$.

It is worth noticing that the experimental conditions to achieve an acceptable reproducibility in the preparation of the photocatalysts are much more critical when $TiCl_4$ is used; in particular by varying slightly the pH to which the preparation of the precursor is carried out, dramatic differences are found in terms of anatase/rutile ratio in the resulting heated polycrystalline solids [30].

We can summarise the above points as follows: (i) different photoactivities are shown by the same semiconductor, namely polycrystalline TiO_2, both in the anatase and in the rutile phases, when different preparation methods are used; (ii) rutile phase obtained from $TiCl_4$ is photoactive when obtained at $T \leqslant 973$ K, inactive when obtained at $T > 973$ K; (iii) rutile phase is obtained from $TiCl_3$ only at higher temperature ($T = 1073$ K) or after more prolonged thermal treatments at 923–973 K and does not show any significant photoactivity; when it is obtained, instead, mixed with anatase at lower temperatures, the photoactivity is the highest within the set of samples prepared from $TiCl_3$; (iv) the above considerations finally can be applied only for photo-oxidation reactions as the photodegradation of organic molecules in aqueous suspension.

Although it is easily understandable why different TiO_2 preparations show different photocatalytic behaviours by taking into account the various physico-chemical and electronic parameters influencing the photoreactivity, it is less straightforward to understand why TiO_2 (rutile) is much photoactive in some and virtually inactive in other cases.

Moreover, some consideration should be given to the enhanced photoactivity of mixed samples in which both rutile and anatase phases are present.

TiO_2 anatase and rutile phases have similar band-gap values (≈ 3.2 and ≈ 3.0 eV, respectively) and, consequently, their different photoactivities cannot be explained by taking into account only these small differences.

On the other hand, the recombination rates of the photoproduced electrons and holes are significantly different for the two phases and, in particular, the rate is higher for rutile [21,31–33]. This parameter could play a negative role for the occurrence of some photoreactions in the case of rutile as, the recombination rate of the pairs being high, they would be less available for the adsorbed reagent species.

Because rutile and anatase are closely juxtaposed when mixed particles are obtained the lifetime of the pairs, and consequently the photoreactivity, could be favoured [34].

The electronic factors, however, are not sufficient to explain the different photoactivities shown by the two phases, and several physico-chemical parameters should be considered, although it is not easy to establish clearly the relative importance of each of them.

Among the physico-chemical parameters, the surface hydroxylation of the catalyst particles is one of the most important; it is well known, indeed, that the major role played by surface hydroxyls as primary oxidant species for photocatalytic reactions [2–5,35–37] and the presence of surface OH groups favours the O_2 adsorption [38–43] and, consequently, its reducibility by photoproduced electrons.

The high photoactivity showed by TiO_2 (rutile) samples obtained at moderate temperatures of firing seems to support the above statement. The moderate temperature of firing do not induce, probably, a dramatic irreversible surface dehydroxylation and the presence of a significant concentration of surface hydroxyls could account for the significant photoactivity shown by these samples. It has been reported [44], moreover, that the surface density of OH groups influence TiO_2 photoactivity, although other parameters can also play a role.

The size of particles in the photocatalyst is another important parameter to be taken into account as the interaction between photons and reacting suspension, in particular the amount of reflected and absorbed photons, depends on it.

When the size of the particles is of the same order of magnitude of the light wavelengths, anelastic scattering occurs; on the other hand, when it is far greater than the light wavelengths the scattering phenomena can be chiefly considered as specular and diffuse reflection processes [45,46].

In both cases it is very difficult to evaluate the fraction of impinging light really absorbed by the reacting system, the geometry of the photoreactor playing also an essential role. Very recently a method to determine experimentally this quantity in a particular set-up and only for aqueous suspensions containing powders whose particles are far greater than the wavelengths of the impinging light has been reported in the literature [47–49] and the smaller particles seem to absorb a more significant fraction of impinging light.

8   INFLUENCE OF DOPANTS ON PHOTOACTIVITY OF TiO$_2$

8.1   METAL IONS

A consequence of the addition of transition metal ions, such as Fe^{3+} or Cr^{3+}, in polycrystalline TiO$_2$ is the increase of the absorption in the visible region, although the band-gap region of TiO$_2$ is often left unaffected [50].

The presence of metal ions determines the formation of a permanent space charge region whose electric force improves the efficiency of the hole–electron separation and consequently the charge transfer from -O-Ti^{4+} to ˙O-Ti^{3+}. By observation of Figure 4.4 it can be noticed that the term C_{sc} would increase and the photoproduced electric charges would be more available for the adsorbed reagents.

From a chemical point of view, moreover, it should be also observed that TiO$_2$ doping is equivalent to the introduction of defects, i.e. Ti^{3+} in the lattice.

By taking into account the above considerations, it is evident that the photo-activity of pure TiO$_2$ can significantly vary when transitional metal ions are present in the lattice and/or on the particle surface, but its dependence is not straightforward. The photo-activity, indeed, can depend on the regime (gas–solid or liquid–solid) under which the photo-process is carried out and the type of reaction (oxidation or reduction) by which the main final product(s) is (are) obtained.

In the following we'll report some hypotheses aiming to explain why the presence of metal ions can be beneficial to, detrimental to or not influential on the photoactivity of TiO$_2$.

First of all we should carefully consider the distribution of the foreign ions, i.e. the lattice and the surface of TiO$_2$ particles. This depends on the amount of the metal ion and on the temperature of firing because some amorphous or micro-crystalline species may segregate on the particle surface if the solubility limit of the used metal ion in TiO$_2$ is exceeded and the diffusion phenomena are related to the temperature.

The degree of dispersion of the metal on the particle surface and the homogeneity of the chemical composition of a single particle depends on the preparation methods and on the firing temperatures to which the various samples have been subjected.

Conventional coprecipitation methods favour a more uniform distribution of the metal species on the TiO$_2$ particle surface in comparison, for instance, with the wet impregnation ones. On the other hand, higher temperatures favour the diffusion of metal ions in the lattice of TiO$_2$ and consequently the formation of a solid solution involving at least a few top layers of TiO$_2$ particles [51–54].

Hence, it is not easy to establish a definite picture for a TiO$_2$-loaded particle as it strongly depends on the amount of metal ion and on its solubility limit in the host oxide. Several bulk and surface techniques are available and have been successfully used to obtain information on the distribution of the metal species in a support.

Among them, X-ray diffraction (XRD), transmission electron microscopy (TEM), scanning electron microscopy (SEM) coupled with electron microprobe used in an energy-dispersive mode (EDX), X-ray photoelectron spectroscopy (XPS), diffuse reflectance spectroscopy (DRS) and Fourier transform infrared spectroscopy (FTIR) have been more widely used.

The presence of Fe^{3+} and Cu^{2+} in TiO_2 (anatase) has been found beneficial up to definite amounts for dinitrogen photoreduction to ammonia [6, 55–57] and CO_2 photoreduction to organic compounds [58–61], respectively, in the presence of water vapour.

Although the mechanistic aspects of the above photoreactions have not yet been completely clarified, due to the very low quantities of products obtained, some hypotheses have been reported to explain the role of the metal ions.

We reported at the beginning of this section that the presence of metal species in the gas–solid regime allows the formation of a permanent space charge region on the particle surface, improving the pairs separation and consequently the photoactivity of pure TiO_2, which is negligible for this kind of reaction. But, amounts of metal ions higher than ca. 2 wt% are detrimental, for instance for the occurrence of ammonia photoproduction, because segregation of metal species can occur during the firing treatments, and these species can partially or totally cover the surface of the TiO_2 particles.

With regard to the photoreduction of CO_2 in the presence of H_2O by using pure and Cu^{2+} doped TiO_2 (anatase) as catalysts CH_4 and CH_3OH are produced, and it has been found that the essential role is played by the presence of photoproduced Cu^+ species for CH_3OH formation. The addition of amounts of Cu^{2+} higher than 3 wt% has a detrimental influence on CH_3OH photoproduction [60]; Ti^{3+} ions, $H^{\cdot}$ and $CH_3^{\cdot}$ radicals have been detected by EPR measurements, but the details of the reaction mechanism are still not clear [10–17, 61].

In a liquid–solid regime in which photo-oxidation of organic pollutants has been carried out, as for instance in aqueous suspensions used for phenol photodegradation, the presence of Fe^{3+} or Cr^{3+} ions in TiO_2 (anatase) has not been found beneficial [50]. Indeed, the energy spectrum of TiO_2 is not changed by doping and the energy of the valence band of anatase, i.e. the energy of the photoproduced holes, is sufficient to oxidise most of the organic substrates [62]. Therefore the above reactions are possible on pure TiO_2 (anatase) from the thermodynamic point of view. The contact between a semiconductor and an electrolyte, indeed, sets a Schottky barrier and the electric field in the depletion layer of the Schottky barrier drives photogenerated electrons and holes in opposite directions if the electrolyte contains a suitable redox couple [63]. The potential drop in the space charge layer developed in the liquid–semiconductor interface (see also Section 4) depends on the difference between the Fermi level of the semiconductor and the Fermi level of the redox electrolyte. The photoreactivity results reported in the literature suggest that the potential drop developed onto the surface of pure TiO_2 (anatase) suspended in water solution containing several kinds of redox couples is sufficient to hasten the hole–electron pair separation, in other words the anatase–electrolyte interface exhibits the essential conditions for the occurrence of the reaction, i.e. a correct oxidation potential and an efficient charge separation. The presence in TiO_2 (anatase) of dopants, such as Cr^{3+} or Fe^{3+}, can induce, of course, a displacement of the Fermi level, but this displacement seems not to affect the effectiveness of the hole–electron separation with respect to the pure oxide. On the other hand, the photoreactivity may be negatively affected by an increased surface recombination of the hole–electron pairs and/or by the variation of the diffusion length of the carriers.

Only very recently there have been reports in the literature [64–66] of the beneficial influence of the presence of tungsten (IV) in polycrystalline TiO_2 (anatase) for the photo-oxidation of 1,4-dichlorobenzene and 4-nitrophenol in aqueous medium.

In this last case the situation is still more complicated as for the first photo-reaction the presence of tungsten was reported to be always beneficial, while for the second one only an amount of tungsten corresponding to ca. 2–3 mol% has been found beneficial.

8.2 PHOTODEPOSITED METALS

In this section some ideas will be briefly presented and the main aims that one can achieve by using semiconductor–metal bifunctional photocatalysts will be high-lighted.

The main conceptual reasons why precious metals are deposited on the surface of TiO_2 particles by using various techniques can be reassumed as follows:

1. The improvement of the separation of the electron–hole pairs whose photo-production is responsible for the increase of the capacitance of the semi-conductor (see Figure 4.4);
2. The increase of the rate of the reduction process due to the presence of a metal efficient by a catalytic point of view (decrease of the term Ψ_{cat} in Figure 4.4).

The necessary but not sufficient requisite in order to achieve the point (1) is related to the work function of the metal: it should be higher than that of the semiconductor.

The setting, however, of an ohmic contact between semiconductor and metal is essential for obtaining their efficient coupling [67].

When this occurs, free electrons produced in the illuminated semiconductor can spontaneously migrate to the phase having a higher work function, i.e. the metal, and thus a more efficient pairs separation is achieved.

The reduction process, consequently, occurs on the metal and the metal itself behaves as a classical catalyst as far as the process under examination is concerned.

Several authors [68–70] have reported a significant increase of the reaction rate for the photocatalytic alcohol's dehydrogenation by using as catalyst polycrystalline TiO_2 (anatase) on whose particles platinum was allowed to deposit.

It is worth noticing that the dehydrogenation of alcohol is not a catalytic process in the presence of pure TiO_2 [71].

For the dehydrogenation of alcohol, the above points (1) and (2) cannot be differentiated when TiO_2 with platinum deposited on its surface is used, as platinum is a specific catalyst for hydrogenation and dehydrogenation processes.

A differentiation, on the other hand, can be done when silver is used instead of platinum as silver does not work as catalyst.

Sclafani *et al.* [72] have observed, indeed, that also in the presence of silver deposited on TiO_2 (Degussa P25 or BDH) the dehydrogenation rate of propan-2-ol increases compared with that observed in the presence of pure TiO_2 in similar experimental conditions. The beneficial influence on the reaction rate can be explained taking into account only the improved pairs separation due to the presence of the metal, as silver is not effective to dissociate hydrogen molecules or to recombine hydrogen atoms but it can attract electrons [73,74].

According to Gerischer *et al.* [75] the reaction rate for photo-oxidation of organic compounds is controlled by the electron transfer rate to O_2 adsorbed on the surface of the catalyst and, when the surface of the semiconductor particles, for instance TiO_2, is modified by the presence of metals belonging to VIII group [76,77], the charge transfer to O_2 and, consequently, the photo-oxidation of the organic compounds become more efficient.

Similar behaviour was previously observed by other authors using single crystal [78] and polycrystalline [79–81] TiO_2 doped with Pt or Pd [80].

Moreover it has been also observed [72] that a significant increase of the photo-oxidation rate of propan-2-ol occurs only for low amount of Pt or Ag and the optimum amount of metal depends on the physico-chemical characteristics of the TiO_2 used and on the reaction ambient.

The above behaviour appears to be general as a critical amount of metal is needed also in other cases [75, 82] in order to obtain a photocatalyst with maximum activity.

A higher concentration of metallic islands on the surface of the particles or an enhancement of their size could negatively influence the photoactivity of the catalyst.

Indeed, a reduced illumination of the surface of the catalyst particles and an increased pairs recombination occurring on the metal surface without involvement of the adsorbed species can occur.

Nevertheless, it has been reported [83] that Pt/TiO_2 is less active than unloaded TiO_2 for the photocatalytic oxidation of neat cyclohexane in the presence of air.

Similar behaviour was observed [84] by using the same catalysts, i.e. Pt/TiO_2 and TiO_2, for the photocatalytic oxidation of 4-chlorophenol in water medium.

These results are in contrast with those obtained by Gerischer and other authors cited above [76–81]. In the following a brief tentative explanation will be provided.

Gerischer hypothesised that the photocatalytic reduction of O_2 on the semiconductor surface is under kinetics control and, consequently, the presence on the surface of TiO_2 of a metal of the VIII group having the double role above described, i.e. to improve the pairs separation and to be efficient by a catalytic point of view, can only give rise to an enhancement of this process.

But, if we consider the possibility of a diffusive control occurring possibly for some samples, the presence of a metal even efficient by a catalytic point of view, should be ineffective or detrimental, depending on its amount.

For higher amounts, the overall process could be slowed down because the pairs recombination could be favoured.

Then, an explanation for this apparent inconsistency among different authors could be found in the different structural and surface characteristics of the TiO_2 samples used as support by Pichat et al. [83,84] and the samples used by other authors [76–81]. For the samples used by Pichat et al., a diffusive control can be hypothesised owing to an high O_2 photoreduction rate. And this is not surprising as TiO_2 Degussa P25, for instance, compared with other TiO_2 samples, showed higher photoadsorption capacity and isotopic exchange rate under UV irradiation [85].

The bifunctional photocatalysts, i.e. powdered semiconductors with a metal deposited on the particle surface, have been employed also for photo-splitting of water [86], for dinitrogen photo-reduction to NH_3 [87], and for CO_2 photo-reduction [88].

9 INFLUENCE OF ADDITIVES ON THE PHOTOACTIVITY OF AQUEOUS SUSPENSIONS OF TiO_2

The photoreactivity can be strongly influenced by the presence of some inorganic species in the reacting system. In the following we'll confine ourselves to outlining some case studies recently reported in the literature concerning the photodegradation of organic pollutants carried out in aqueous suspensions containing polycrystalline TiO_2 of both phases (anatase and rutile).

The main aim is to explain briefly why the presence of some inorganic species such as Ag^+, Fe^{3+}, Fe^{2+}, and H_2O_2 strongly influences the photoreactivity of TiO_2.

Previously, we have reported that the separation of the photoproduced pairs on the catalyst surface is an essential step influencing the photoreactivity. And moreover, the negligible photoreactivity shown by dehydroxylated TiO_2 rutile [29] could be related to the scarce role played by oxygen as electron trap, O_2 adsorption on the catalyst surface strongly depending on the hydroxyl concentration [41].

By having in mind the above considerations, the behaviour of most of the reacting aqueous TiO_2 suspensions in which O_2 or He are bubbled and containing inorganic species in addition to organic molecules, can be interpreted.

We begin to discuss the role played by Ag^+ ions in the photodegradation of phenol, used as 'probe' reaction.

The presence of Ag^+ ions in the reacting system is, of course, only of academic interest, but its utilisation allows one to derive some useful information.

Ag^+ behaves as an efficient electron scavenger because its adsorption, followed by the charge transfer process, significantly occurs [24,89] on both anatase and rutile TiO_2 phases, according to the following reaction:

$$Ag^+ + e_{cb}^- \rightarrow Ag_{surf} \qquad (4.7)$$

Consequently, the difference of photoactivity shown by polycrystalline TiO_2 (anatase) and TiO_2 (rutile) for photodegradation of organic compounds in aqueous suspensions in the presence of O_2 becomes insignificant when a suitable concentration of Ag^+ ions, i.e. *ca.* 10^{-2} mol l^{-1} for dispersions containing 10^{-3} mol l^{-1} of phenol [24,89], is used. The comparable photoactivity of the two phases can be explained by hypothesising that the Ag^+ reductive process on rutile occurs faster than the O_2 reduction process, thus rendering the holes more available for the photo-oxidation processes. A further evidence of the above insight is the similar photoactivity shown also by the two phases in the presence of Ag^+ ions when O_2 is changed with He.

A similar result is obtained when H_2O_2 instead of Ag^+ is added in the absence of O_2 to the reaction system. Indeed, the photocatalytic behaviour of TiO_2 (rutile) is not much different, in this case, from that of TiO_2 (anatase) [89]. The rutile photoactivity can be justified taking into account the previous considerations, i.e. the oxidation of phenol in the presence of rutile significantly occurs when an electron scavenger (H_2O_2 in this case) kinetically more effective over rutile than O_2 is added.

With regard to the addition of Fe^{3+} ions, the situation is more complicated. The role played by Fe^{3+} ions is quite different from that of Ag^+ and H_2O_2. These last species appeared to be more efficient than O_2 to trap the photoproduced electrons both on the anatase and rutile particles. Consequently, they were used to demonstrate that the negligible photoactivity of rutile phase was mainly due to kinetic constraints related to the photoreduction of O_2.

Fe^{3+} ions can induce, on the other hand, an enhancement of the O_2 photoreduction rate by participating in some step(s) of the reaction mechanism.

It is well known, indeed, that O_2 photoadsorption followed by photoreduction gives rise to formation of peroxo species on the surface of TiO_2 particles and these species hinder the subsequent reduction. Therefore their decomposition could be an important kinetic step beneficially influencing the overall process of O_2 photoreduction. Fe^{3+} ions that can be reduced to Fe^{2+} by the photoproduced electrons can play an essential role in the decomposition of surface peroxo species. The reaction between Fe^{2+} and H_2O_2 has been well known, on the other hand, in homogeneous medium for many years (Fenton reagent).

Photoreactivity has been found to be strongly dependent on the concentration of these ions in heterogeneous systems. No beneficial influence on photoreactivity was shown for either rutile or anatase phases in the presence of $5 \times 10^{-2} \, mol \, l^{-1} \, Fe^{3+}$ ions and He, for suspensions containing $10^{-3} \, mol \, l^{-1}$ of phenol [89,90].

When O_2, instead of He, was used, the presence of Fe^{3+} ions was ineffective for rutile and detrimental for anatase.

A plausible explanation for this behaviour may be the strong absorption of most of the impinging photons by the solution in which Fe^{3+} is dissolved, the absorbing properties of the medium playing in this case a major role, although in addition a parasite role of Fe^{3+} ions (their reduction and subsequent oxidation) cannot be excluded.

When the concentration of Fe^{3+} is reduced, the photoactivity increases in the presence of O_2 and TiO_2 anatase, a maximum of photoreactivity being shown for $[Fe^{3+}] = 5 \times 10^{-4} \, mol \, l^{-1}$ (Figure 4.5).

If the role of Fe^{3+} was only to absorb photons as a filter, then photoreactivity, by decreasing the Fe^{3+} concentration, should increase to the level observed in the

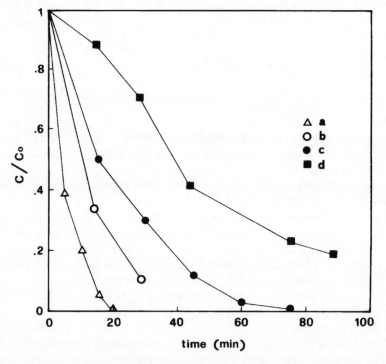

Figure 4.5. C/C_o versus reaction time where C is the phenol concentration at time t and C_o is the initial phenol concentration. TiO_2 (anatase) in the presence of oxygen and various Fe^{3+} concentrations: $5 \times 10^{-4} \, M$ (curve a); $5 \times 10^{-5} \, M$ (curve b); $0 \, M$ (curve c); $5 \times 10^{-3} \, M$ (curve d). $C_o = 10^{-3} \, M$

absence of Fe^{3+}. The presence of a photoreactivity maximum and the large range of Fe^{3+} concentration $(1 \times 10^{-3}–1 \times 10^{-5} \, mol \, l^{-1})$ observed for which the photoreactivity is higher than in the absence of Fe^{3+} suggest that, when suitable amounts of Fe^{3+} ions are present, some processes having a beneficial influence on the photoreaction can occur.

Among them, as above anticipated, the most relevant are the following:

$$Fe^{3+} + e^-_{(cb)} \rightarrow Fe^{2+} \tag{4.8}$$

$$Fe^{2+} + HO^{\cdot}_{2(surf)} + H^+_{(aq)} \rightarrow Fe^{3+} + H_2O_{2(surf)} \tag{4.9}$$

$$Fe^{2+} + H_2O_{2(surf)} + H^+_{(aq)} \rightarrow Fe^{3+} + OH^{\cdot} + H_2O \tag{4.10}$$

$$H_2O_{2(surf)} + e^-_{(cb)} \rightarrow OH^- + OH^{\cdot} \tag{4.11}$$

In reactions (4.8)–(4.10), Fe^{3+} behaves as cocatalyst and mediator.

Fe^{2+} ions produced according to reaction (4.8) react with photo-produced H_2O_2 and/or peroxide radicals according to reactions (4.9) and (4.10). These last species can be photoproduced on the surface of the catalyst [42–43,91] according to the following reactions:

$$O_{2(ads)} + e^-_{cb} + H^+_{(aq)} \rightarrow HO^{\cdot}_{2(surf)} \tag{4.12}$$

$$HO^{\cdot}_{2(surf)} + e^-_{cb} + H^+_{(aq)} \rightarrow H_2O_{2(surf)} \tag{4.13}$$

By taking into account all the reactions described above, the concentration of $OH^{\cdot}$ radicals, i.e. the primary oxidant species of the photodegradation process, increases and experimental conditions for the continuous photoproduction of Fenton reactive $(H_2O_2 + Fe^{2+})$ are achieved.

Further indirect evidence for the occurrence of the mechanism described above is given by the net decrease of the photodegradation rate in the presence of $[Fe^{3+}] = 5 \times 10^{-4} \, mol \, l^{-1}$ (the optimum concentration) and He, instead of O_2. In this case, indeed, reaction (4.8) can still take place but H_2O_2 is not photoproduced (in the absence of O_2) and consequently reactions (4.9) and (4.10) do not occur.

The negligible photoreactivity found for rutile in the presence of $[Fe^{3+}] = 5 \times 10^{-4} \, mol \, l^{-1}$ and oxygen can be justified in a similar way.

In this last case, probably, H_2O_2 and $HO^{\cdot}_2$ are not significantly photoproduced on the catalyst surface because adsorption of O_2 is not high, and consequently Fe^{3+} cannot play any beneficial role.

Fe^{2+} ions showed a beneficial influence on the photodegradation of organic pollutants similarly to Fe^{3+} ions, under the same experimental conditions, as they can directly react with H_2O_2 and $HO^{\cdot}_{2(surf)}$ according to reactions (4.9) and (4.10).

To complete this section some considerations on the addition of H_2O_2 will be done. The use of hydrogen peroxide and UV radiation or Fe^{2+} ions (Fenton reagent) for decontaminating water effluents is well known in the literature in homogeneous media [92–95]. In these cases, the UV radiation or the presence of Fe^{2+} give rise to the production of $OH^{\cdot}$ radicals from H_2O_2 according to the following reactions:

$$H_2O_2 \xrightarrow{h\nu} 2OH^{\cdot} \qquad (4.14)$$

$$H_2O_2 + Fe^{2+} \longrightarrow OH^- + OH^{\cdot} + Fe^{3+} \qquad (4.15)$$

It is, however, worth noticing that only very energetic photons with $\lambda \leqslant 254\,nm$ are needed for the occurrence of reaction (4.14). Moreover, both reactions (4.14) and (4.15) in several cases do not allow one to achieve a complete mineralisation of the organic molecules present in the polluted water effluent, but only their partial (photo)-oxidation.

When H_2O_2 is added, instead, to a heterogeneous aqueous system containing polycrystalline TiO_2 and organic pollutants, its presence is beneficial for the photodegradation of these molecules, both in the presence of He and O_2 [96–100].

The explanation for this behaviour is again the production of $OH^{\cdot}$ radicals and their high probability of interaction with organic adsorbed species. The mechanism, indeed, in this case is different from that occurring in homogeneous systems as the production of $OH^{\cdot}$ radicals is due to the interaction between adsorbed H_2O_2 molecules and photoproduced electrons (see equation (4.11)).

ACKNOWLEDGEMENTS

The authors wish to thank MURST (Ministero dell'Università e della Ricerca Scientifica e Tecnologica), Rome for financial support.

REFERENCES

[1] *Photoelectrochemistry, Photocatalysis and Photoreactors, Fundamentals and Developments*, M. Schiavello (Ed.), Reidel, Dordrecht, 1985.
[2] *Homogeneous and Heterogeneous Photocatalysis*, E. Pelizzetti and N. Serpone (Eds.), Reidel, Dordrecht, 1986.
[3] *Photocatalysis and Environment. Trends and Applications*, M. Schiavello (Ed.), Kluwer, Dordrecht, 1988.
[4] *Photocatalysis. Fundamentals and Applications*, E. Pelizzetti and N. Serpone (Eds.), Wiley, New York, 1989.
[5] *Photocatalytic Purification and Treatment of Water and Air*, D.F. Ollis and H. Al Ekabi (Eds.), Elsevier, Amsterdam, 1993
[6] N. Schrauzer and T.D. Guth, *J. Am. Chem. Soc.* **99** (1977), 7189.
[7] V. Augugliaro, A. Lauricella, L. Rizzuti, M. Schiavello and A. Sclafani, *Int. J. Hydrogen Energy* **7** (1982), 845.

[8]   M. Schiavello, L. Rizzuti, R.I. Bickley, J.A. Navio and P.L. Yue, *Proc. 8th Int. Congr. Catal.* **3** (1984), 383.
[9]   J. Soria, J.C. Conesa, V. Augugliaro, L. Palmisano, M. Schiavello and A. Sclafani, *J. Phys. Chem.* **95** (1991), 274, and references therein.
[10] M. Halmann, *Nature* **275** (1978), 115.
[11] C. Hemminger, R. Carr and G.A. Somorjai, *Chem. Phys. Lett.* **57** (1978), 100.
[12] B.A.-Blajeni, M. Halmann and J. Hanassen, *Solar Energy* **25** (1980), 165.
[13] M. Ulman, A.H.A. Tinnemans, A. Mackor, B. Aurian-Blajeni and M. Halmann, *Int. J. Solar Energy* **1** (1982), 213.
[14] M. Anpo, *Res. Chem. Intermed.* **11** (1989), 67.
[15] M. Anpo and K. Chiba, *Proc. VII SHHC* 17–21 May 1992, Tokyo, p.80.
[16] H. Sakurai, S. Tsubota and M. Haruta, *Appl. Catal. A: General* **102** (1993), 125.
[17] H. Yamashita, H. Nishiguchi, N. Kamada, M. Anpo, Y. Teraoka, H. Hatano, S. Ehara, K. Kikui, L. Palmisano, A. Sclafani, M. Schiavello and M.A. Fox, *Res. Chem. Intermed.* **20** (1994), 815.
[18] A. Fujishima and K. Honda, *Nature* (London), **238** (1972), 37.
[19] T. Ohnishi, Y. Nakato and H. Tsubomura, *Ber. Buns. Phys. Chem.* **79** (1975), 523.
[20] A.A. Krasnovsky and G.P. Brin in *Photosynthetic Oxygen Evolution*, H. Metzner (Ed.), Academic Press, New York, 1978, pp.405–410.
[21] K. Okamoto, Y. Yamamoto, H. Tanaka, M. Tanaka and A. Itaya, *Bull. Chem. Soc. Jpn.* **58** (1985), 2015.
[22] R.W. Matthews, *J. Catal.* **111** (1988), 264.
[23] V. Augugliaro, L. Palmisano, A. Sclafani, C. Minero and E. Pelizzetti, *Toxicol. Environ. Chem.* **16** (1988), 89.
[24] A. Sclafani, L. Palmisano and E. Davì, *New J. Chem.* **14** (1990), 265.
[25] V. Augugliaro, E. Davì, L. Palmisano, M. Schiavello and A. Sclafani, *Appl. Catal.* **65** (1990), 101 and references therein.
[26] V. Augugliaro, L. Palmisano, M. Schiavello, A. Sclafani, L. Marchese, G. Martra and F. Miano, *Appl. Catal.* **69** (1991), 323.
[27] H. Hidaka, J. Zhao, E. Pelizzetti and N. Serpone, *J. Phys. Chem.* **96** (1992), 2226.
[28] S.J. Teichner and M. Formenti in *Photoelectrochemistry, Photocatalysis and Photoreactors*, M. Schiavello (Ed.), Reidel, Dordrecht, 1985, p.457.
[29] A. Sclafani, L. Palmisano and M. Schiavello, *J. Phys. Chem.* **94** (1990), 829.
[30] F. Cavani, E. Foresti, F. Parrinello and F. Trifirò, *Appl. Catal.* **38** (1988), 311.
[31] H. Kawaguchi and T. Uejima, *Kogaku Kogaku Ronbunshu* **9** (1983), 107.
[32] H. Kawaguchi, *Environ. Technol. Lett.* **5** (1984), 471.
[33] K.M. Schindler and M. Kunst, *J. Phys. Chem.* **94** (1990), 8222 and references therein.
[34] R.I. Bickley, T. Gonzalez-Carreño, J.S. Lees, L. Palmisano and R.J.D. Tilley, *J. Solid State Chem.* **92** (1991), 178.
[35] C.S. Turchi and D.F. Ollis, *J. Catal.* **122** (1990), 178.
[36] D. Lawless, N. Serpone and D. Meisel, *J. Phys. Chem.* **95** (1991), 5166.
[37] J.C. D'Oliveira, G. Al-Sayed and P. Pichat, *Environ. Sci. Technol.* **24** (1990), 990.
[38] M. Primet, P. Pichat and M.V. Mathieu, *J. Phys. Chem.* **75** (1971), 1216.
[39] R.I. Bickley and F.S. Stone, *J. Catal.* **31** (1973), 389.
[40] R.I. Bickley and R.K. Jayanty, *J. Chem. Soc. Faraday Discuss.* **58** (1974), 194.
[41] A.H. Boonstra and C.A.H.A. Mutsaers, *J. Phys. Chem.* **79** (1975), 1694.
[42] G. Munuera, V. Rives-Arnau and A. Saucedo, *J. Chem. Soc. Faraday Trans.* **1**, **75** (1979), 736.
[43] G. Munuera, A.R. Gonzalez-Elipe, V. Rives-Arnau, A. Navio, P. Malet, J. Soria, J.C. Conesa and J. Sanz in *Adsorption and Catalysis on Oxide Surfaces*, M. Che and G.C. Bond (Eds.), Elsevier, Amsterdam, 1985, p.113.

[44] R. Campostrini, G. Carturan, L. Palmisano, M. Schiavello and A. Sclafani, *Mater. Chem. Phys.* **38** (1994), 277.

[45] R. Siegel and J.R. Howell in *Thermal Radiation Heat Transfer*, B.J. Clark and D. Damstra (Eds.), McGraw-Hill, New York, 1972, Chap. 20, p. 661.

[46] H.C. Hottel and A.F. Sarofim, *Radiative Transfer*, McGraw-Hill, New York, 1967.

[47] M. Schiavello, V. Augugliaro and L. Palmisano, *J. Catal.* **127** (1991), 332.

[48] V. Augugliaro, L. Palmisano and M. Schiavello, *AIChe J.* **37** (1991), 1096.

[49] V. Augugliaro, V. Loddo, L. Palmisano and M. Schiavello, *J. Catal.* **153** (1995), 32.

[50] L. Palmisano, V. Augugliaro, A. Sclafani and M. Schiavello, *J. Phys. Chem.* **92** (1988), 6710.

[51] R.I. Bickley, T. Gonzalez-Carreño and L. Palmisano, *Mater. Chem. Phys.* **29** (1991), 475.

[52] C. Martin, I. Martin, V. Rives, L. Palmisano and M. Schiavello, *J. Catal.* **134** (1992), 434.

[53] L. Palmisano, M. Schiavello, A. Sclafani, C. Martin, I. Martin and V. Rives, *Catal. Letters* **24** (1994), 303.

[54] R.I. Bickley, T.G. Gonzalez-Carreño, A.R. Gonzalez-Elipé, G. Munuera and L. Palmisano, *J. Chem. Soc. Faraday Trans.* **90** (1994), 2257.

[55] M.I. Litter and J.A. Navio, *J. Photochem. Photobiol. A: Chem.* **98** (1996), 171.

[56] E. Endoh, J.K. Leland and A.J. Bard, *J. Phys. Chem.* **90** (1986), 6223.

[57] V. Augugliaro, F. D'Alba, L. Rizzuti, M. Schiavello and A. Sclafani, *Int. J. Hydrogen Energy* **7** (1982), 851.

[58] T. Inoue, A. Fujishima, S. Konishi and K. Honda, *Nature* **277** (1979), 637.

[59] M. Halmann, V. Katzir, E. Borgarello and J. Kiwi, *Sol. Energy* **10** (1984), 85.

[60] M. Anpo and K. Chiba, *J. Mol. Catal.* **207** (1992), 74.

[61] M. Halmann, M. Ulman and B.A.-Blajeni, *Solar Energy* **31** (1983), 429.

[62] M.A. Fox, *J. Am. Chem. Soc.* **16** (1983), 314.

[63] H. Gerischer, *J. Electroanal. Chem.* **58** (1975), 263.

[64] W. Lee, W-M. Gao, K. Dwight and A. Wold, *Mater. Res. Bull.* **27** (1992), 685.

[65] Y.R. Do, W. Lee, K. Dwight and A. Wold, *J. Solid State* **108** (1994), 198.

[66] G. Marcì, L. Palmisano, A. Sclafani, A.M. Venezia, R. Campostrini, G. Carturan, C. Martin, I. Martin and V. Rives, *J. Chem Soc. Faraday Trans.,* **92** (1996), 819.

[67] R. Memming in *Photoelectrochemistry, Photocatalysis and Photoreactors*, M. Schiavello (Ed.), Reidel, Dordrecht, 1985, p.107.

[68] P. Pichat, J.M. Herrmann, J. Disdier, H. Courbon and M.N. Mozzanega, *Nouv. J. Chim.* **5** (1981), 627.

[69] S. Nishimoto, B. Ohtani and T. Kagiya, *J. Chem. Soc. Faraday Trans. I* **81** (1985), 21.

[70] P. Pichat, *ACS Symp. Ser.* **278** (1985), 21.

[71] J.M. Herrmann, J. Disdier and P. Pichat, *J. Catal.* **113** (1988), 72.

[72] A. Sclafani, M.N. Mozzanega and P. Pichat, *J. Photochem. Photobiol. A: Chem.* **59** (1991), 181.

[73] (a) A. Henglein, *J. Phys. Chem.* **83** (1979), 2558; (b) A. Henglein, *Ber. Bunsenges. Phys. Chem.* **84** (1980), 254.

[74] P.C. Lee and D. Meisel, *J. Catal.* **70** (1981), 160.

[75] C. M. Wang, A. Heller and H. Gerischer, *J. Am. Chem. Soc.* **114** (1992), 5230.

[76] H. Gerischer and A. Heller, *J. Phys. Chem.* **95** (1991), 5261.

[77] H. Gerischer and A. Heller, *J. Electrochem. Soc.* **139** (1992), 113.

[78] T. Kobayashi, H. Yoneyama and H. Tamura, *J. Electrochem. Soc.* **103** (1983), 1706.

[79] B. Kräutler and A.J. Bard, *J. Am. Chem. Soc.* **100** (1978), 5985.

[80] I. Izumi, W.W. Dumm, K.O. Wilbourn, F.F. Fan and A.J. Bard, *J. Phys. Chem.* **84** (1980), 3207.

[81] I. Izumi, F.F. Fan and A.J. Bard, *J. Phys. Chem.* **85** (1981), 218.

[82] A. Wold, *Chem. Mater.* **5** (1993), 280.

[83] W. Mu, J.-M. Herrmann and P. Pichat, *Catal. Lett.* **3** (1989), 73.

[84] G. Al-Sayyed, J.C. D'Oliveira and P. Pichat, *J. Photochem. Photobiol. A: Chem.* **58** (1991), 99.

[85] H. Courbon, J.-M. Herrmann and P. Pichat, *J. Phys. Chem.* **88** (1984), 5210.

[86] (a) S. Sato and J.M. White, *Chem. Phys. Lett.* **72** (1980), 83; (b) K. Yamaguchi and S. Sato, *J. Chem. Soc. Faraday Trans.* **81** (1965), 1237; (c) D. Duonghong, E. Borgarello and M. Grätzel, *J. Am. Chem. Soc.* **103** (1982), 4685; (d) S. Nakabayashi, A. Fujishima and K. Honda, *Chem. Phys. Lett.* **102** (1983), 464.

[87] N.N. Rao, S. Dube, Manjubala and P. Natarajan, *Appl. Catal. B: Environ.* **5** (1994), 33.

[88] (a) J. C. Hemminger, R. Carr and G. A. Somorjai, *Chem. Phys. Lett.* **57** (1978), 100; (b) O. Ishitani, C. Inoue, Y. Suzuki and T. Ibusuki, *J. Photochem. Photobiol. A: Chem.* **72** (1993), 269; (c) V. Heleg and I. Willner, *J. Chem. Soc. Chem. Commun.* (1994), 2113.

[89] A. Sclafani, L. Palmisano and E. Davì, *J. Photochem. Photobiol. A: Chem.* **56** (1991), 113.

[90] A. Sclafani and L. Palmisano, *Gazz. Chim. Ital.* **120** (1990), 599.

[91] R.I. Bickley in *Specialist Periodical Reports of the Royal Society of Chemistry: Chemical Physics of Solids and their Surfaces*, London, 1978, Vol.7, p.118.

[92] P.C. Ho, *J. Environ. Sci. Technol.* **20** (1986), 260.

[93] *Management of Hazardous and Toxic Wastes in the Process Industries*, S.T. Kolaczkowski and B.D. Crittenden (Eds.), Elsevier, London, 1987.

[94] M.B. Borup and E.J. Middlebrooks, *Water Sci. Technol.* **19** (1987), 381.

[95] B.A. Weir, D.W. Sundstrom and H.E. Klei, *Hazard. Waste Hazard. Mat.* **4** (1987), 165.

[96] V. Augugliaro, E. Davì, L. Palmisano, M. Schiavello and A. Sclafani, in *Environmental Contamination*, A.A. Orio (Ed.), CEP Consultants, Edinburgh, 1988, p.206.

[97] K. Tanaka, T. Hisanaga and K. Harada, *New J. Chem.* **13** (1989), 5.

[98] K. Tanaka, T. Hisanaga and K. Harada, *J. Photochem. Photobiol. A: Chem.* **48** (1989), 155.

[99] V. Augugliaro, E. Davì, L. Palmisano, M. Schiavello and A. Sclafani, *Appl. Catal.* **65** (1990), 101.

[100] E.J. Wolfrum and D.F. Ollis in *Aquatic and Surface Photochemistry*, G.R. Helz, R.G. Zepp and D.G. Crosby (Eds.), Lewis Publishers, Boca Raton, U.S.A., 1994, p.451.

5 Photocatalytic Reductions— Photocatalytic Reduction of Carbon Dioxide with Water and Hydrogenation of Unsaturated Hydrocarbons with Water

M. ANPO and H. YAMASHITA

Department of Applied Chemistry, Osaka Prefecture University Gakuen-cho 1-1, Sakai, Osaka 593, Japan

1 INTRODUCTION

The most significant as well as large-scale chemical reaction on Earth is natural plant photosynthesis which produces glucose and oxygen from carbon dioxide and water. This phenomenal, life-sustaining reaction has attracted much attention in recent years as a model for the development of an artificial reaction system for the storage and conversion of solar energy to useful chemical energy [1–3]. On the other hand, the large-scale emission of carbon dioxide into the atmosphere has

Heterogeneous Photocatalysis, Edited by M. Schiavello
© 1997 John Wiley & Sons Ltd.

wrought one of the most serious and dangerous problems upon the Earth, especially with regard to the devastating consequences of the greenhouse effect [4–5].

The reduction and/or fixation of carbon dioxide can be said to be one of the most important areas of research in chemistry today, not only for solving the many urgent problems resulting from the pollution of the global environment but also for finding ways to maintain vital carbon resources which are being depleted by the burning of fossil fuels as well as making the design of an artificial photosynthesis reaction system possible [6].

Furthermore, such research will make sense only if the reactions can be effectively and efficiently achieved using a natural energy source—the most safe, clean, and ideal being solar energy. The utilisation of solar energy for the reduction and/or fixation of carbon dioxide can be made possible by considering the following two reaction systems: (i) the photocatalytic reduction and/or fixation of CO_2 with H_2O into CO, HCOOH, CH_3OH, and CH_4, etc. using reactive photocatalysts such as small particle powdered TiO_2 semiconductors and, (ii) the photoelectrochemical reduction and/or fixation of CO_2 in aqueous systems using semiconducting electrodes.

In this chapter, we will focus on the photocatalytic reduction and/or fixation of CO_2 with H_2O using heterogeneous photocatalytic systems, one of the most desired as well as challenging goals we face in this field. Studies have been carried out by many researchers on extremely small TiO_2 particles and on highly dispersed anchored titanium oxide catalysts as well as on fine particle metal sulfide catalysts, under UV irradiation [7–38]. Although the efficiency of the photocatalytic reduction and/or fixation of CO_2 on the catalytic surfaces is still too low to realise the efficient control of CO_2 emissions and the supply of carbon resources, investigations leading to a complete understanding of the fundamental mechanisms behind the reaction will be instrumental in applying the principles for improving the efficiency and selectivity on a large, global scale.

With these objectives of research in mind, in this chapter, we have summarised our recent results on the characteristic features of the photocatalytic reduction and/ or fixation of CO_2 with H_2O on various types of active titanium oxide catalysts and metal sulfide catalysts. In addition, special attention has been focused on the relationship between the local structure of the active sites on the catalysts and their photocatalytic reactivities. Investigations on the dynamic properties of the excited states of the active sites as well as on the direct detection of the reaction intermediate species have also been carried out in order to gain an understanding of the photocatalytic reaction mechanisms at the molecular level.

2   THE PHOTOCATALYTIC REDUCTION OF CO_2 WITH H_2O

How efficient the photosynthetic reduction of CO_2 with H_2O to produce CH_4 and CH_3OH is will depend strongly upon the photocatalyst used [7–38]. Among photocatalysts, the semiconducting TiO_2 catalyst is the most reactive and stable so

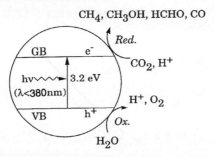

CH$_4$, CH$_3$OH, HCHO, CO

Figure 5.1. Reaction scheme for the photocatalytic reduction of CO$_2$ with H$_2$O on bulk TiO$_2$

that many investigations have been carried out using semiconducting TiO$_2$ photocatalysts. In Figure 5.1, the primary processes for the photocatalytic reduction and/or fixation of CO$_2$ with H$_2$O on semiconducting TiO$_2$ photocatalysts has been illustrated. However, as shown in Table 5.1, the electrochemical reduction potentials of CO$_2$ in these reactions are quite high as compared to the electron energy level in the TiO$_2$ photocatalyst.

When the particle size of the semiconducting TiO$_2$ is decreased, the band gap between the conduction band and the valence band becomes larger, making it suitable and applicable for the reduction of CO$_2$ [39–41]. For extremely fine particle TiO$_2$ photocatalysts less than 100 Å of the particle, the size quantum effect and/or the effects of the surface modification in its coordination geometry plays a significant role in the reactivity of the photocatalyst [39–47]. As a result, the electrons and holes which are produced by UV irradiation within the ultrafine particles of TiO$_2$ and the highly dispersed titanium oxide species exhibit more unique and high reactivities than for those produced in large particle TiO$_2$ photocatalysts.

Table 5.1. Thermodynamic data for CO$_2$ conversion to other C$_1$ molecules

$CO_2(g) + 2H_2O(l) \rightarrow HCOOH(aq) + \frac{1}{2}O_2(g)$	$\Delta G = 1.428\,eV$
$CO_2(g) + H_2O(l) \rightarrow HCHO(aq) + O_2(g)$	$\Delta G = 1.350\,eV$
$CO_2(g) + 2H_2O(l) \rightarrow CH_3OH(aq) + 3/2O_2(g)$	$\Delta G = 1.119\,eV$
$CO_2(g) + 2H_2O(l) \rightarrow CH_4(aq) + 2O_2(g)$	$\Delta G = 1.037\,eV$
$CO_2 + e^- \rightarrow CO_2^-$	$E^\circ = -1.9\,V$
$CO_2 + 2H^+ + 2e^- \rightarrow HCO_2H$	$E^\circ = -0.61\,V$
$CO_2 + 2H^+ + 2e^- \rightarrow CO + H_2O$	$E^\circ = -0.53\,V$
$CO_2 + 4H^+ + 4e^- \rightarrow C + 2H_2O$	$E^\circ = -0.20\,V$
$CO_2 + 4H^+ + 4e^- \rightarrow HCHO + H_2O$	$E^\circ = -0.48\,V$
$CO_2 + 6H^+ + 6e^- \rightarrow CH_3OH + H_2O$	$E^\circ = -0.38\,V$
$CO_2 + 8H^+ + 8e^- \rightarrow CH_4 + 2H_2O$	$E^\circ = -0.24\,V$

2.1 PHOTOCATALYTIC REACTIONS ON VARIOUS TITANIUM
OXIDE CATALYSTS

2.1.1 Liquid–Solid Reaction Systems

This section deals with the photocatalytic reduction of CO_2 with H_2O to produce CH_4 and CH_3OH on semiconductor photocatalysts, especially on TiO_2 suspended in an aqueous solution.

Inoue *et al.* [9] first reported that HCOOH, HCHO, and a trace amount of CH_3OH are produced by the reduction of CO_2 with water under the illumination of aqueous suspension systems involving a variety of semiconductor powders such as TiO_2 and $SrTiO_3$ as can be seen in Table 5.2. Such photocatalytic production of CH_4 from CO_2 and H_2O has also been reported by Hemminger *et al.* on the $Pt/SrTiO_3$ catalyst [7,8], while pioneering works on the photoreduction of CO_2 on semiconductors were summarised by Halmann [11] and by Aresta *et al.* [16]. However, when water was used as the reductant, the efficiency of CO_2 reduction was low and a detailed characterisation of this reaction has yet to be clarified. Although several researchers have made efforts to enhance the efficiency and selectivity for the photoreduction of CO_2, few significant advances have been reported.

Hirano *et al.* have observed CH_3OH as the main photoreduction product from CO_2 when a Cu-containing TiO_2 suspension system was irradiated [35]. The addition of Hg or Pt onto the TiO_2 photocatalyst accelerated the formation of HCHO and the carbon species, respectively. Recently, Ishitani *et al.* [36] have reported that the deposition of metals (Pd, Pt, Rh, etc) on the TiO_2 photocatalyst considerably accelerated the photocatalytic reduction of CO_2 to CH_4 and CH_3CO_2H. For the photocatalytic reduction of CO_2 in suspended aqueous systems, the Pd/TiO_2 system was found to exhibit a very high selectivity for the production

Table 5.2. The photocatalytic reduction of CO_2 with H_2O on the various types of powdered semiconductors in aqueous suspensions

Photocatalysts	Production rate ($\mu mol\,h^{-1}$)			Efficiency (%)
	HCOOH	HCHO	CH_3OH	
(a) $SrTiO_3$	1.2	0.05	0.02	0.022
(b) TiO_2	0.71	0.014	—	0.011
(c) TiO_2	2.3	0.066	0.022	0.039
(d) $BaTiO_3$	0.11	0.012	0.001	0.003
(e) $BaTiO_3$	0.23	0.027	—	0.005
(f) $LiNbO_3$	—	0.029	—	0.001
(g) $LiNbO_3$	1.56	0.15	—	0.030

Pretreatments: (a) heated 6.5 h at 1100 °C under vacuum; (b) 7 h at 1000 °C under vacuum; (c) 2 h at 500 °C under vacuum; (d) 22 h at 1000 °C under $H_2 + Ar$ (1 : 1); (e) 4 h at 700 °C in air, with 5×10^{-3} mol% Nd_2O_3; (f) 2 h at 1150 °C in air; (g) 2.5 h at 1150 °C in air, with 0.5 mol% Nd_2O_3.

of CH_4. RuO_2 dispersed on a TiO_2 has been used by Grätzel and coworkers for the reduction of CO_2 to CH_4. High yields were obtained especially when transition metal complexes were used as photosensitisers [23,24].

Titanates, such as $SrTiO_3$, that show a conduction band at a potential of about 0.2 V more negative than TiO_2, have also attracted a great deal of interest. They appear to afford the best chemical conversion of the absorbed photons and this is explained also in terms of a strong absorption of carbon dioxide [14,15]. The addition of Rh, Ir, and Pt metal oxides also affects the selectivity. HCOOH was the predominant product and the addition of IrO_x was the most effective for HCOOH formation [15]. Raphael and Malati have reported that UV irradiation of aqueous carbonate solutions in the presence of Pt-doped titanate powders led to the formation of carbon as the main reduction product accompanied by HCHO [19].

2.1.2 Gas–Solid Reaction Systems

2.1.2.1 Small particle TiO_2 catalysts

Our investigations of the photocatalytic reduction of CO_2 with H_2O to produce CH_4 and CH_3OH on small particle TiO_2 photocatalysts in the gas–solid reaction systems will now be discussed [26–29].

UV irradiation of the powdered TiO_2 catalysts in the presence of a gaseous mixture of CO_2 and H_2O led to the evolution of CH_4 into the gas phase at 275 K. Trace amounts of C_2H_4 and C_2H_6 were also produced. The yields of these products increased with the UV irradiation time, while no products were detected under dark conditions. The CH_4 yield was almost zero in the reaction of CO_2 without H_2O and increased when the amount of H_2O was increased. These results suggest that for powdered TiO_2 catalysts, the photocatalytic reduction of CO_2 to produce CH_4 and C_2-compounds from CO_2 and H_2O takes place photocatalytically in the solid–gas phase systems. The formation of CH_4 as the main product at 343 K has also been recently observed by Saladin et al., and the formation of partially reduced $TiO_{2-\delta}$ by the photoreduction of TiO_2 is considered to be the active species [22].

The photocatalytic reduction of CO_2 with H_2O has been carried out on four different types of TiO_2 catalysts, and the yields of CH_4 formation are shown in Table 5.3. The level of photocatalytic reactivity, based on the CH_4 yields, was found to depend on the type of TiO_2 catalyst, in the order of JRC-TIO-$4 > -5 > -2 > -3$. The photocatalytic reactivities of these four TiO_2 catalysts were also investigated for the hydrogenation reaction of methyl acetylene with H_2O [48] and for the isomerisation of 2-butene [48] and these findings are also shown in Table 5.3. The tendency for catalytic activity in the photocatalytic reduction of CO_2 with H_2O is in a manner similar to those for the hydrogenation and isomerisation of olefins, proving that the reduction of CO_2 with H_2O occurs photocatalytically over the powdered TiO_2 catalyst.

Table 5.3. Physical properties and photocatalytic activity of standard TiO_2 catalysts

Catalyst (JRC-TIO-)	Surface area ($m^2 g^{-1}$)	CO_2 ads. ($\mu mol\,g^{-1}$)	Acid conc. ($\mu mol\,g^{-1}$)	Relative -OH conc.	Band gap (eV)	Reduction[a] of CO_2 ($\mu mol\,h^{-1}\,g^{-1}$)[d]	Hydrogenation[b] of olefins ($\mu mol\,h^{-1}\,g^{-1}$)[d]	Isomerisation[c] of olefins ($\mu mol\,h^{-1}\,g^{-1}$)[d]
2 (anat.)	16	1	6	1	3.47	0.03	0.20	2.5
3 (ruti.)	51	17	22	1.6	3.32	0.02	0.12	1.0
4 (anat.)	49	10	5	3.0	3.50	0.17	8.33	9.4
5 (ruti.)	3	0.4	7	3.1	3.09	0.04	0.45	3.8

[a] CH_4 yield in reaction of CO_2 with H_2O.
[b] Conversion in reaction of methyl acetylene with H_2O.
[c] Conversion in reaction of 2-butene.
[d] As reported previously, there was no correlation between activity of the catalysts and their surface areas.

The various chemical and physical properties of these four TiO_2 catalysts can also be seen in Table 5.3. From these results, it is likely that the anatase-type TiO_2 which has a large band gap and numerous surface OH groups is preferable for efficient photocatalytic reactions. The band gap increase is accompanied by a shift in the conduction band edge to higher energy levels. This shift causes the reductive potential to shift to more negative values which in turn causes a great enhancement in the photocatalytic reactivity. This effect seems to be more crucial for difficult or slow reactions such as the reduction of CO_2 with H_2O. Concerning the role of the surface OH groups, the surface OH groups and/or physisorbed H_2O play a significant role in photocatalytic reactions *via* the formation of OH radicals and H˙ radicals. Thus, the data shown in Table 5.3 can be interpreted in terms of the small changes in the band gap and/or the concentration of the surface OH groups.

Figure 5.2 shows the ESR signals obtained under UV irradiation of the anatase-type TiO_2 (JRC-TIO-4) catalyst in the presence of CO_2 and H_2O at 77 K. The ESR signals are attributed to the characteristic photogenerated Ti^{3+} ions ($g_\perp = 1.9723$ and $g_\parallel = 1.9628$) and H˙ radicals (with 490 G splitting), as well as ˙CH_3 radicals which consist of four lines with intensity ratios of $1:3:3:1$, having a hyperfine splitting of $H\alpha = 19.2$ G and g-value of 2.002. The signal intensity of ˙CH_3 radicals decreased on increasing the amount of H_2O, indicating that CH_3 radicals react easily to form CH_4 in the presence of enough H_2O. These results clearly suggest that ˙CH_3 radicals are the intermediate species and react with H˙ radicals that are formed by the reduction of protons (H^+) supplied from H_2O adsorbed on the catalyst.

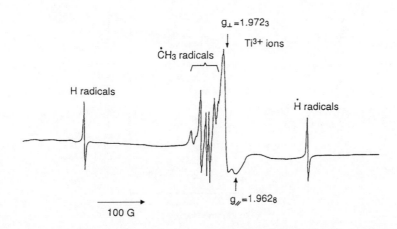

Figure 5.2. ESR signals obtained with TiO_2 under UV irradiation in the presence of CO_2 and H_2O at 77 K (CO_2: 0.12 mmol, H_2O: 0.37 mmol, irradiation time: 6 hr)

2.1.2.2   Cu-loaded and Pt-loaded TiO_2 catalysts

The effect of Cu-loading on the photocatalytic reduction of CO_2 with H_2O on TiO_2 catalysts was investigated in our laboratory. Although the loading of Cu onto the small particle TiO_2, i.e. the Cu/TiO_2 photocatalyst, led to a suppression of the CH_4 yield, a new formation of CH_3OH could be observed as shown in Figure 5.3. The Cu/TiO_2 catalysts having different amounts of Cu-loading, i.e. 0.3–1.0 Cu wt%, exhibited the photocatalytic reactivity to produce CH_3OH with a yield of 0.006–0.004 μmol.g-cat^{-1}.

In the XPS spectra of the Cu/TiO_2 catalyst just before the photo-reaction, the absence of the satellite peak in the $Cu(2p_{3/2,1/2})$ spectra and the position of the $Cu(L_3VV)$ peak suggest that the main species of copper in the catalyst is Cu^+. Recently, it has been reported that Cu^+ catalysts play a significant role in the photoelectrochemical production of CH_3OH from the CO_2 and H_2O system [49] and the results obtained in the present study correspond well with these reports. Following the photocatalytic reduction, the Cu/TiO_2 catalyst exhibited a new peak at 299 eV in the C(1s) XPS spectra, suggesting that carboxylate groups are formed on the catalyst during the photocatalytic reaction. These carboxylate groups which accumulated on the catalyst may be the primary intermediate species for the photocatalytic reduction of CO_2 with H_2O.

The effect of Pt-loading on the TiO_2 catalyst was also investigated. The yield of CH_4 increased remarkably when the amount of Pt added was increased (0.1–1.0 wt%). On the other hand, the addition of excess Pt as well as the excess loading of Cu onto the TiO_2 catalyst were undesirable for an efficient reaction to occur.

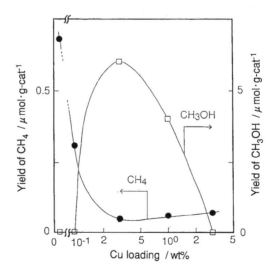

Figure 5.3.   Effect of Cu^{2+} loading on CH_4 and CH_3OH yields in photocatalytic reduction of CO_2 with H_2O on Cu/TiO_2 catalyst (CO_2: 71 μmol, H_2O: 71 μmol, reaction time: 4 h)

Regarding the reaction intermediates, Solymosi *et al.* have investigated the effect of UV irradiation on the activation of CO_2 on oxidised, evacuated or reduced TiO_2 and Rh/TiO_2 powdered catalysts by means of FT-IR spectroscopy [50–52]. They observed an appearance of the IR bands at around 1640 and 1219 cm^{-1} due to the CO_2^- species in bent form under UV irradiation of the Ph/TiO_2 photocatalyst in the presence of CO_2. This observation suggests that an electron transfer from the irradiated catalyst to the adsorbed CO_2 takes place, resulting in the formation of a CO_2^- anion.

They have also observed that on Rh/TiO_2, the photo-induced dissociation of CO_2 occurred, as indicated by the development of CO bands at around 2020–2040 cm^{-1} due to the adsorbed CO. From these IR results, it has been assumed that the added CO_2 molecules on the Rh/TiO_2 catalyst interacts with the surface through both C (linked to Rh atoms) and O atoms (linked to the oxygen vacancy of titania) and that the charge transfer from the catalyst to the CO_2 molecules induced by UV irradiation leads to the C—O bond cleavage. This proposal indicates that the CO_2^- species may be the key species in the photochemical reactions of CO_2 using metal loaded TiO_2 semiconductor photocatalysts.

2.1.2.3 *TiO₂ single crystals*

With a well-defined catalyst surface such as a single crystal, detailed information on the reaction mechanism can be obtained at the molecular level [53–55]. Therefore, the photocatalytic reduction of CO_2 with H_2O on rutile-type $TiO_2(100)$ and $TiO_2(110)$ single crystal surfaces and the reaction intermediates formed on the surfaces has been detected using high-resolution electron energy loss spectroscopy (HREELS), a highly sensitive means of investigation [30].

UV irradiation of the $TiO_2(100)$ single crystal catalyst in the presence of a mixture of CO_2 and H_2O led to the evolution of CH_4 and CH_3OH in the gas phase at 275 K, whereas only CH_3OH yields was detected with the $TiO_2(110)$ single crystal catalyst. The CH_4 and CH_3OH on the two types of TiO_2 single crystals are shown in Table 5.4. It is clear that the efficiency and selectivity of the photo-catalytic reactions strongly depend on the type of TiO_2 single crystal surface. The yield of CH_3OH formation is much higher on $TiO_2(100)$ than on $TiO_2(110)$, while the formation of CH_4 is only observed on $TiO_2(100)$ and not on $TiO_2(110)$.

It is likely that the photo-formed electrons localise on the surface sites of the excited TiO_2 to play a significant role in the photoreduction of CO_2 molecules into intermediate carbon species [26–29]. The surface Ti atoms may act as an electron moiety on the surfaces, i.e. a reductive site. According to the surface geometric models for $TiO_2(100)$ and $TiO_2(110)$, the atomic ratio (Ti/O) of the top-surface Ti and O atoms which have geometric spaces large enough to have direct contact with CO_2 and H_2O molecules is higher on $TiO_2(100)$ than on $TiO_2(110)$ surface. In the excited state, the surface with a higher Ti/O surface ratio, i.e. $TiO_2(100)$, exhibits a more reductive tendency than $TiO_2(110)$. Such a reductive surface allows a more

Table 5.4. Yields of the formation of CH_4 and CH_3OH
in the photocatalytic reduction of CO_2 with H_2O at 275 K

Single crystal	Yields CH_4	$(nmol\,h^{-1}\,g\text{-cat}^{-1})$ CH_3OH
TiO_2 (100)	3.5	2.4
TiO_2 (110)	0	0.8

CO_2 (124 μmol g^{-1}), H_2O (372 μmol g^{-1})

facile reduction of CO_2 molecules especially for the formation of CH_4 which was found to be a vital step in the photoreduction of CO_2 with H_2O on the TiO_2 catalyst.

Figure 5.4 shows a typical HREELS spectrum of a clean TiO_2(100) single crystal surface before and after UV irradiation in the presence of CO_2 and gaseous H_2O, respectively. Following UV irradiation, two peaks were detected at around 2920 and 3630 cm^{-1} in addition to the peaks attributed to the optical phonon of the TiO_2 lattice at around 680 cm^{-1}, which could be assigned to the C–H stretching vibration of the CH_x species [56] and to the O–H stretching of the surface hydroxyl groups, respectively [57]. On the other hand, without UV irradiation, only a weak peak assigned to the O–H stretching vibration was observed by the exposure of the catalyst to reactant gases, indicating that UV irradiation is indispensable for the reduction of CO_2 molecules and the formation of active H and CH_x species.

2.1.2.4   Highly dispersed anchored titanium oxide catalysts

As with the study on extremely small TiO_2 particles, the investigations on highly dispersed anchored titanium oxides as photocatalysts have attracted a great deal of attention [58–67]. With well-defined highly dispersed catalysts, it is possible not only to achieve more active and selective photocatalytic systems but also to obtain detailed information on the nature of the active sites and the reaction mechanisms at the molecular level [1].

The titanium oxide anchored onto Vycor glass was prepared using a facile reaction of $TiCl_4$ with the surface OH groups on the transparent porous Vycor glass (Corning code 7930) in the gas phase at 453–473 K, followed by treatment with H_2O vapour to hydrolyse the anchored compound. A preparation scheme of the anchored catalyst is shown in Figure 5.5.

Figure 5.6(a) shows a typical photoluminescence spectrum of the titanium oxide anchored onto Vycor glass at 77 K. Excitation by light at around 280 nm brought about an electron transfer from the oxygen to titanium ion, resulting in the formation of pairs of the trapped hole centre (O$^-$) and an electron centre (Ti^{3+}) [1, 25, 58, 63]. The observed photoluminescence is attributed to the radiative decay process from the charge transfer excited state of the titanium oxide moieties having a tetrahedral coordination to their ground state. As shown in Figure 5.6(b–e), the

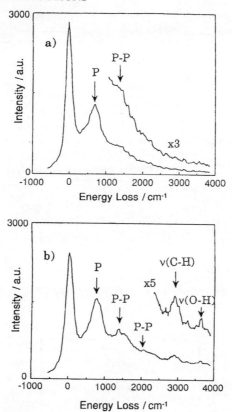

Figure 5.4. HREEL spectra of $TiO_2(100)$ before and after UV irradiation: (a) clean surface; (b) after UV irradiation for 5 min in the presence of CO_2 (1 torr) and gaseous H_2O (3 torr) (P; phonon peak, P-P; phonon–phonon peak)

addition of H_2O or CO_2 molecules at 275 K onto the anchored titanium oxide leads to the efficient quenching of the photoluminescence with differing efficiencies. Such an efficient quenching with CO_2 or H_2O suggests not only that the emitting sites are highly dispersed on the surfaces, but also that the added CO_2 or H_2O interacts and/or reacts with the anchored titanium oxide species in its excited states. Because the addition of CO_2 led to a less effective quenching than with the addition of H_2O, the interaction of the emitting sites with CO_2 was weaker with H_2O.

Figure 5.7 shows the XANES and FT-EXAFS spectra of the titanium oxide catalysts anchored onto Vycor glass and bulk TiO_2 (anatase). Although the bulk TiO_2 exhibits three small pre-edge peaks in its XANES spectrum, the anchored titanium oxide catalyst (Figure 5.7(a)) exhibits only a single and intense pre-edge peak which is similar to that of tetra-*iso*-propoxy titanium, having a tetrahedral coordination, indicating the presence of tetrahedrally coordinated titanium oxide

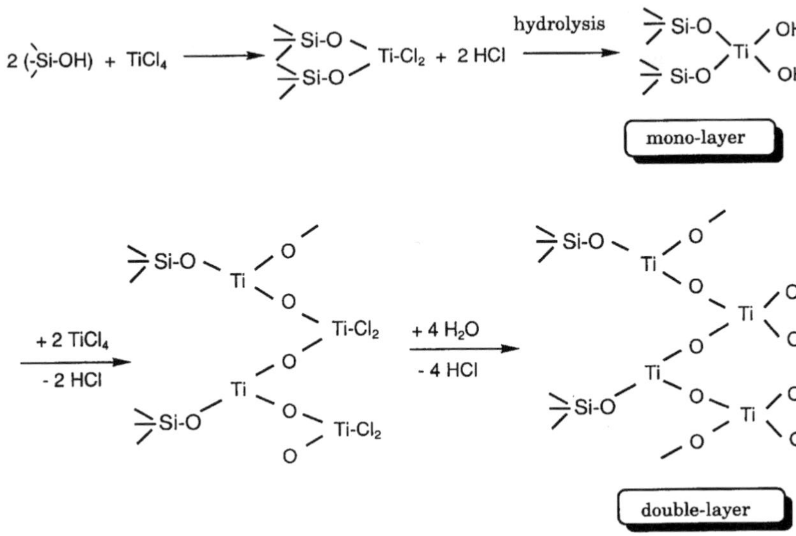

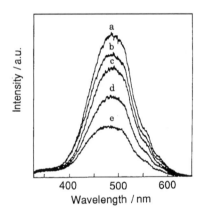

Figure 5.5. Scheme for preparation of highly dispersed titanium oxide catalyst by an anchoring method

Figure 5.6. Photoluminescence spectrum of the anchored titanium oxide at 77 K (a), and the quenching by added (b) CO_2 and (c–e) H_2O. Excitation beam: 280 nm; amount of added CO_2: (b) 0.6 mmol g^{-1}; amounts of added H_2O: (c) 0.04, (d) 0.08, (e) 0.2 mmolg^{-1}

species on the surfaces [63]. The FT-EXAFS spectrum of the anchored titanium oxide catalyst (Figure 5.7(a)) exhibits only Ti-O peaks, indicating the presence of an isolated titanium oxide species. These findings indicate that highly dispersed isolated tetrahedral titanium oxide species are formed on Vycor glass by an anchoring method.

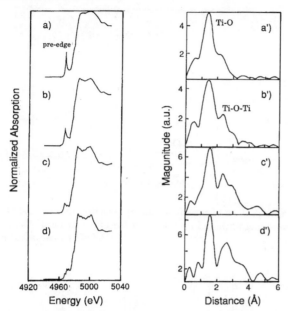

Figure 5.7. Ti K-edge XANES spectra (left) and FT-EXAFS spectra (right) of the titanium oxide catalysts anchored on Vycor glass with (Ti-O) layer (a, a'), three layers (b, b'), 5 layers (c, c') catalysts, and bulk TiO_2 (anatase) (d, d')

UV irradiation of the anchored titanium oxide catalysts in the presence of a mixture of CO_2 and H_2O led to the evolution of CH_4, CH_3OH, and CO in the gas phase at 323 K, as well as trace amounts of C_2H_4 and C_2H_6. The evolution of small amounts of O_2 was also observed. However, no products could be detected under dark conditions. Figure 5.8 shows the time profiles for the production of CH_4 in relation to the UV irradiation time. The yields of the products are in a good linear relationship with the irradiation time, for long periods up to 6–8 h, and then decline gradually, especially after prolonged UV irradiation. The total CH_4, CH_3OH and CO yields were larger under UV irradiation at 323 K than at 275 K. It is thus clear that these photocatalytic reactions proceed more efficiently at higher temperatures. Figure 5.8 also shows the effect of the H_2O/CO_2 ratio on the photocatalytic reactivity. As shown in Figure 5.8, the efficiency of the photocatalytic reaction strongly depends on the ratio of H_2O/CO_2 and its reactivity increases with an increase in the H_2O/CO_2 ratio; however, an excess amount of H_2O suppresses the reaction rates.

Figure 5.9 shows the effect of the number of anchored Ti-O layers on the absorption edge of the catalysts and the efficiency of the photocatalytic reactions as well as the relative yields of the photoluminescence [28]. It was found that only catalysts with highly dispersed monolayer titanium oxide layers exhibit high photocatalytic reactivity and a photoluminescence spectrum at around 480–500 nm.

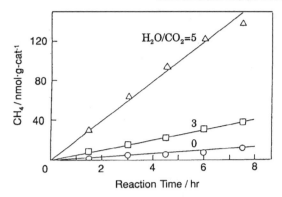

Figure 5.8. The time profiles of the photocatalytic reaction of CO_2 and H_2O to produce CH_4 on the titanium oxide anchored on PVG and the effect of the H_2O/CO_2 ratio on the yields of the products. (Numbers represent the ratio of H_2O/CO_2, CO_2: $0.05\,mmol\,g^{-1}$.)

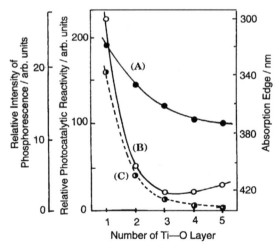

Figure 5.9. The effects of the number of the Ti-O layers of the anchored titanium oxide catalysts on the absorption edge of the catalysts (A), the reaction yields (B), and the relative yields of the photoluminescence

Only the tetrahedral titanium oxide species exhibit a photoluminescence spectrum when it is excited at around 300 nm. These findings clearly suggest that the tetrahedrally coordinated titanium oxide species act as active photocatalysts for the reduction of CO_2 with H_2O.

Figure 5.10 shows the effect of the support on the distribution (selectivity) and yields of the photoproducts on the anchored titanium oxide catalysts [26, 31]. There

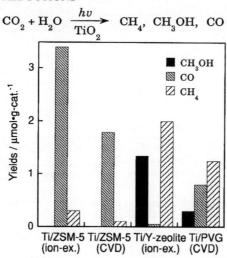

Figure 5.10. The products distribution of the photocatalytic reduction of CO_2 with H_2O on various kinds of anchored titanium oxide catalysts. (CO_2: $0.12\,mmol\,g^{-1}$, H_2O: $0.60\,mmol\,g^{-1}$.)

is a remarkable difference in the reactivity and selectivity among the catalysts. CO is formed as the main product on the titanium oxide anchored on ZSM-5, whereas the titanium oxides anchored on the Y-zeolite and PVG produce CH_4 as the main product accompanied by the formation of CH_3OH. Titanium oxide anchored on the Y-zeolite exhibited the highest photocatalytic activity as well as selectivity for the formation of CH_4 and CH_3OH.

In order to obtain a detailed analysis of the mechanisms behind the photocatalytic reaction, various mixtures of the reactant gases, i.e., CO and H_2O, CO_2 and H_2, and CO_2 alone were used. In the case of CO_2 alone, no reaction occurred on any of the catalysts, indicating that the presence of H_2O is indispensable for the reaction. On the Ti/ZSM-5 catalyst, CO was the main product of the CO_2 and H_2O mixture; however, no reaction took place from the CO and H_2O system. On the other hand, on the Ti/Y-zeolite and Ti/PVG catalysts, CH_4 and CH_3OH were mainly produced from the CO_2 and H_2O mixture as well as from the CO and H_2O mixture. These results indicate that CO is an intermediate species in the reaction of the CO_2 and H_2O mixture. However, on the Ti/ZSM-5 catalyst CO easily desorbs from the surface, resulting in no further reaction for the formation of CH_4 and CH_3OH.

UV irradiation of the anchored titanium oxide catalysts in the presence of CO_2 and H_2O at 77 K led to the appearance of ESR signals attributed to the C˙ radicals, H˙ atoms and Ti^{3+} ions. After the disappearance of these ESR signals at around 275 K, the formation of CH_4 and CH_3OH could be observed. At 77 K, the ESR

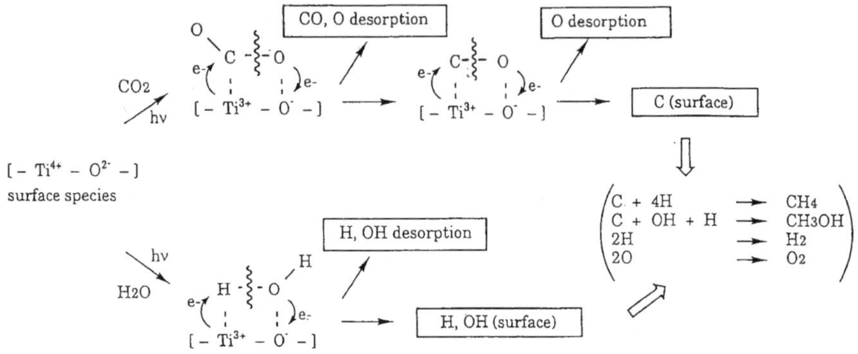

Figure 5.11.   Schematic representation of photocatalytic reduction of CO_2 with H_2O on the anchored titanium oxide

signals assigned to the O_2^- species was observed in the reaction system. The formation of graphitic carbon species was also observed by XPS studies of the anchored titanium oxide catalyst after the photoreaction. It can, therefore, be said that the graphitic carbon species and O_2^- species are formed and stimulated on the surface, which may be responsible for the decrease in catalytic activity after prolonged UV irradiation.

Figure 5.11 shows the reaction mechanism of the photocatalytic reduction of CO_2 with H_2O on the anchored titanium oxide catalyst. CO_2 and H_2O molecules interact with the excited state of the photoinduced $(Ti^{3+}-O^-)^*$ species and the reduction of CO_2 and the decomposition of H_2O proceed competitively, depending on the ratio of H_2O/CO_2. These interactions result in the formation of H· atoms, ·OH radicals and carbon species, and these intermediate radical species react with each other to form CH_4 and CH_3OH.

2.1.2.5   Ti/Si binary oxide catalysts

Ti/Si binary oxides involving different Ti contents were prepared by the sol–gel method using mixtures of tetraethylorthosilicate and titanium-*iso*-propoxide. Ti/Si binary oxides with only a small Ti content exhibited the photoluminescence spectra at around 480 nm upon excitation at around 280–300 nm [26, 28, 31, 67]. Decreasing the Ti content in the binary oxides led to an increase in the photoluminescence intensity while its peak wavelength shifted to shorter wavelength regions.

Through the findings mentioned above, it was found that the photoluminescence spectrum and photocatalytic reactivity of the highly dispersed titanium oxides can be observed only with catalysts having tetrahedrally coordinated Ti (IV) ions in the oxides. In fact, the analysis of the Ti K edge XANES spectra and FT-EXAFS spectra of the Ti/Si binary oxides indicated that titanium ions are located in a tetrahedral coordination environment in the SiO_2 matrices. The XAFS (XANES and FT-EXAFS) spectra of the catalysts having higher contents of Ti ions no longer exhibited the characteristic spectrum and suggest that titanium ions are present in an octahedral coordination. The ESR measurement of the Ti^{3+} ions produced by the UV irradiation of the catalyst at 77 K in the presence of H_2 also supported that the titanium oxide species in the Ti/Si binary oxides having a small amount of Ti ions are present in a highly dispersed tetrahedral coordination in the SiO_2 matrices.

UV irradiation of the Ti/Si binary oxide catalysts in the presence of a gaseous mixture of CO_2 and H_2O was found to lead to the reduction of CO_2 to produce CH_4 and CH_3OH as the main products. As shown in Figure 5.12, a parallel relationship between the specific photocatalytic activities of the titanium oxide species and the photoluminescence yields of the Ti/Si binary oxides clearly indicates that the appearance of high photocatalytic activity for the binary oxides is closely associated with the formation and reactivity of the charge transfer excited complex of the highly dispersed tetrahedral titanium oxide species.

The XAFS, ESR and photoluminescence investigations of the Ti/Si binary oxide catalysts indicated that these catalysts prepared by the sol–gel method are able to sustain a tetrahedral coordination of titanium oxide species until the Ti content reaches approximately up to 20 wt% as TiO_2. Thus, we can see that the Ti/Si binary oxides having a high Ti content (up to 20 wt%) can be successfully utilised as active photocatalysts for the efficient reduction of CO_2 with H_2O in the gas–solid system at 323 K.

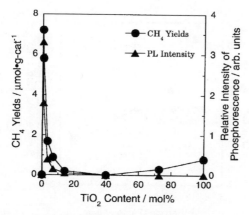

Figure 5.12. Effects of TiO_2 contents on the yields of CH_4 and the relative intensity of photoluminescence of the Ti/Si binary oxide catalysts at 77 K. (CO_2: 0.07 mmol g^{-1}, H_2O: 0.33 mmol g^{-1}, 325 K.)

2.2 PHOTOCATALYTIC REACTIONS ON NON-OXIDE CATALYSTS

2.2.1 Metal Sulfide Catalysts

Recent studies have revealed that quantised semiconductor particles are effective photocatalysts since their band gap is wide and the recombination rate of a photoformed electron–hole pair is relatively slow in such ultrafine particles [68–74]. Several systems for the photoreduction of CO_2 were recently developed applying colloidal chalcogenide semiconductors such as ZnS, CdS and their mixtures as photocatalysts in the presence of different sacrificial electron donors (hole scavengers) [75–83]. In this case, as shown in Table 5.1, two-electron reduction products such as CO and formate (HCO_2^-), and four- and six-electron reduction products such as HCHO and CH_3OH were produced.

However, the quantum efficiencies of these photoreductions were very low, except for the case of small ZnS microcrystallites in the presence of highly reactive hole scavengers. Henglein and Gutierrez first reported the efficient photoreduction of CO_2 into formic acid using SiO_2-stabilised and quantised ZnS as the photocatalyst in the presence of 2-propanol as the sacrificial electron donor [68]. The high quantum yield for the formation of formic acid has been reported at 0.8.

Recently, using chalcogenide microcrystal semiconductors with various types of hole scavengers, Yoneyama *et al.* have carried out the photocatalytic reduction of CO_2 with water to form CO, formate (HCO_2^-) and CH_3OH [76–79]. It was found that their yields and selectivity depended a great deal on the type and particle size of the semiconductors as well as the type of scavengers used. ZnS microcrystallites (3–5 nm) prepared by a precipitation method with different molar ratios of Zn^{2+} to S^{2-} exhibited different photoactivities for CO_2 reduction [76]. As shown in Figure 5.13 the smaller the particle size of the ZnS microcrystallites and the greater the

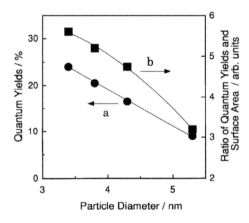

Figure 5.13. The quantum yields for the CO_2 photoreduction to formate and the ratio of the quantum yields and surface area as a function of the size of ZnS microcrystallites

molar ratio of the surface Zn^{2+} to S^{2-}, the higher the quantum efficiency using 2-propanol as the hole scavenger was, indicating that size quantised ZnS is very active for the photoreduction of CO_2 into the formate species.

The loading of several kinds of metal ions and their effect on the ZnS micro-crystallites have also been investigated [78]. It has been reported that the addition of Cd ions was the most effective in enhancing the quantum efficiency for the photoreduction of CO_2 into formate species. The Cd-loaded ZnS prepared by photodeposition of $Cd(ClO_4)_2$ on the ZnS colloid exhibited reactivity twice that of the unloaded ZnS photocatalyst. However, the solid solutions of microcrystals of ZnS–CdS were found to be sensitive to visible light although the photocatalytic reactivity was not enhanced.

The types of stabilisers in this photoreaction system were also investigated [77]. The use of stabilisers having negative charges such as polyacrylic acid for CdS microcrystallites in aqueous solutions selectively yields a format species, while positively charged stabilisers such as polyethyleneimine worked well for the selective production of CO. This difference in the yields of formates and CO is considered to come from the difference in the configuration of adsorbed CO_2. The CO_2^- formed by a one-electron reduction of adsorbed CO_2 caused an end-on type adsorption with the negatively charged stabilisers resulting in the production of formate while causing a side-on adsorption with the positively charged stabilisers resulting in the production of CO.

Yanagida *et al.* have demonstrated that ZnS microcrystallites (2–5 nm) catalyse the quantitative photoreduction of aliphatic ketones into alcohols using Na_2S and Na_2SO_3 as the sacrificial electron donors under UV irradiation [80]. The quantised ZnS microcrystallites contribute to an efficient photoinduced charge separation and electron transfer while the coexistence of S^{2-} and SO_3^{2-} ions cooperatively suppress the formation of the electron trap sites associated with photocorrosion, that is, the formation of lattice defects and impurities. The use of ZnS microcrystallites with a low density of surface defects can provide novel photocatalysts due to the stabilisation of the electron–hole pair against the recombination while eliminating the deactivation of electrons by the surface traps.

The application of these defect-free ZnS microcrystallites for the efficient photoreduction of CO_2 was achieved at pH 7 by using the quantised ZnS in the presence of HS^- which was taken from Na_2S and NaH_2PO_2 as the electron donor [81, 82]. A large quantity of formate (HCO_2^-) accompanied by a small quantity of CO was formed, and a simultaneous evolution of a large amount of H_2 was observed in this photoreduction reaction. An increase in the content of Na_2S brought about an increase in the yield of formate accompanied by a decrease in H_2 which results from the suppression of surface defects as electron trap sites, promoting the evolution of H_2.

On the other hand, the ultrafine ZnS crystallites (< 2 nm) prepared in an organic solvent exhibited higher efficiency for CO_2 photoreduction to form formate and CO in the presence of triethylamine as the electron donor [83]. In this system, the addition of excess Zn^+ resulted in an increase in the yields of CO and H_2 products accompanied by a suppression of formate production. Because the addition of excess Zn^+ caused the formation of surface defects (S^{2-} vacancies), it is estimated that the selective formation of formate proceeds on defect-free surfaces and CO formation proceeds efficiently on the surfaces with many defect sites. These results suggest that the product distribution depends a great deal on the differences in the configuration of adsorbed CO_2 and this could be attributed to whether surface defects were present or not. Using the semi-empirical quantum chemical calculation, the configuration of adsorbed CO_2 on the defect-free and also defect-containing surfaces were estimated and the reaction scheme for CO_2 photoreduction was proposed as shown in Figure 5.14.

By the combination of the ZnS microcrystallites and enzymes (methanol dehydrogenase, MDH) with assistance from the electron mediator (pyrroloquinoline quinone, PQQ), Yoneyama et al. have developed a novel route for the photoreduction of CO_2 to CH_3OH [79]. In this system, ZnS microcrystallites were used

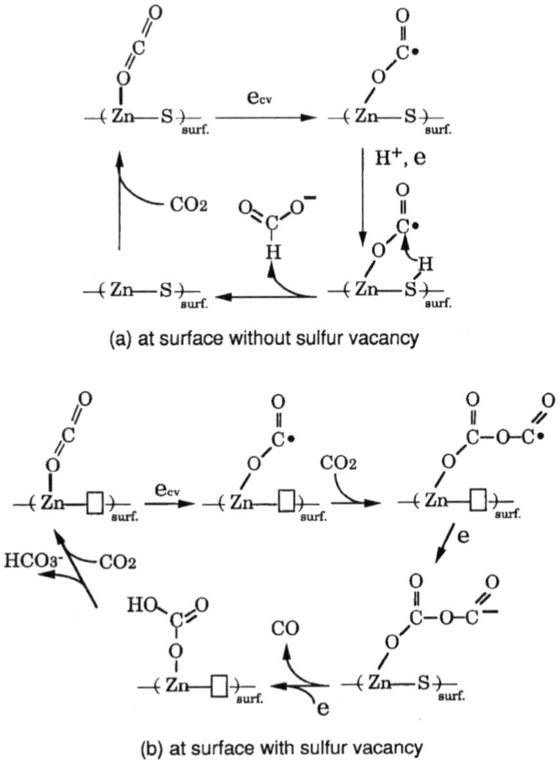

(a) at surface without sulfur vacancy

(b) at surface with sulfur vacancy

Figure 5.14. Schematic representation of photocatalytic reduction of CO_2 on the ZnS microcrystallites: (a) at surface without sulphur vacancy; (b) at surface with sulphur vacancy

as a photocatalyst to reduce CO_2 to produce formate and the resulting formate was in turn reduced by the use of MDH and PQQ.

2.2.2 Photocatalytic Reaction on Organic Electron Transport Sensitisers

CO_2 photoreduction in the presence of organic electron transport sensitisation has been investigated [84–91] and it was found that with linear aromatic chain molecules such as poly(p-phenylenes) (PPP) and oligo(p-phenylenes)(OPP-n) in the presence of triethylamine as a sacrificial electron donor, a series of novel systems in which photoreduction of CO_2 to form formate (HCO_2^-) and a small amount of CO could be achieved under UV irradiation [86–90]. Among OPP-n, p-terpheny (OPP-3) and p-quanterphenyl (OPP-4) showed high photocatalytic activity for the formation of formate, in which the quantum yields for OPP-3 and Opp-4 were 0.072 and 0.084, respectively [86, 87].

Laser flash photolysis and pulse radiolysis studies revealed that the photocatalysis initially starts from the reductive quenching of the singlet state of OPP-n by TEA followed by the formation of the radical anion of OPP-n (OPP-n$^{\cdot-}$), resulting in the direct electron transfer from Opp-n$^{\cdot-}$ to CO_2. Although in a solvent of methanol–acetonitrile, OPP-3 was easily degraded under UV irradiation, by combining with the Co-cyclam as an electron mediator, the efficient photoreduction of CO_2 to CO and formate was able to proceed [88–90]. It was also found that the nitrogen-containing photosensitiser (Phenazine) with efficient electron and hydrogen mediators such as the Co-cyclam is more effective in producing formate selectively [91]. Although the design of photosensitised systems using heteroaromatic ring compounds is a promising idea for the design of artificial photosynthesis systems, the stability of organic compounds needs to be improved.

3 PHOTOCATALYTIC HYDROGENATION OF UNSATURATED HYDROCARBON WITH H_2O

Boonstra and Mutsaers have found that the photocatalytic hydrogenation of alkenes and alkynes occurs over TiO_2 containing adsorbed water, and have proposed that H atoms produced according to the reaction Ti-OH $+ h\nu \rightarrow$ Ti-O* $+$ H* are responsible for the hydrogenation [92]. However, as has also been shown by Schrauzer and Guth the observed reaction is not a simple hydrogenation and the reaction products undergo a fission of the carbon–carbon bond [93]. Such a fission of the bond cannot be explained solely by the concept that the reaction is caused by H atoms. It has also been suggested that the reaction could be related to the photodecomposition of water.

The hydrogen atoms produced by the photodecomposition of water adsorbed on the TiO_2 surface may play an important role in the hydrogenation reactions. The participation of the photocatalytic decomposition of water in the photo-hydrogenation of ethylene was also demonstrated by Sato and White, using Pt/TiO_2 powders [94]. We have proposed that positive holes generated by UV irradiation participate in the reaction and the charge transfer complex of transition metal oxides, such as the Mo-oxide species, plays a significant role in the fission of the $C=C$ bond of alkenes [95]. In the present study, the nature of the bond fission in the photocatalytic hydrogenation over TiO_2 has been clarified. Such information appears to be necessary for a more complete understanding of the photocatalytic reactions over TiO_2 caused by the active species formed from water.

3.1   PHOTOCATALYTIC HYDROGENATION ON TiO_2

Table 5.5. shows the photo-formed products and their yields obtained with TiO_2, which contains a sufficient amount of water to form a monolayer (wet TiO_2 catalyst). With alkynes, the major photo-formed products are alkanes formed by hydrogenation accompanied by the fission of the carbon–carbon bond of the reactant molecules (hereafter referred to as photohydrogenolysis). With alkenes and alkynes, in addition to such products undergoing the bond fission, alkanes formed without carbon–carbon bond fission (hereafter referred as to photohydrogenation) are produced as minor products. In addition to these hydrogenated products, oxygen-containing compounds such as CH_3CHO as well as CO and CO_2 were detected. Similar results were obtained with TiO_2, which does not contain a sufficient amount of water to form a monolayer. Although it was seen that alkanes were again formed by hydrogenation accompanied by photohydrogenolysis, their yields were much lower than those obtained with wet TiO_2 [96, 97].

When the degassing temperature of the wet TiO_2 was increased the yields of photohydrogenolysis decreases drastically, approaching about 9% of the initial value at 473 K as shown in Figure 5.15. It is known that TiO_2 outgassed at room temperature causes the adsorbed water to exist in two forms, i.e., physisorbed

Table 5.5.   Yields of photo-formed products from various unsaturated hydrocarbons on TiO_2 at 300 K

Reactants	Products (μmol)		
$CH{\equiv}CH$	CH_4 (1.65),	C_2H_4 (0.112),	n-C_4H_8 (0.378),
$CH_3{-}C{\equiv}CH$	C_2H_6 (4.86),	CH_4 (0.451),	C_3H_6 (0.312)
$C_2H_5{-}C{\equiv}CH$	C_3H_8 (4.65),	CH_4 (1.10),	C_4H_6 (0.380),
	C_2H_6 (0.147),	$CH_3{-}CH{=}C{=}CH_2$ (0.295)	
$CH_2{=}C{=}CH_2$	C_2H_6 (2.27),	CH_4 (0.023),	C_4H_{10} (0.436),
	C_3H_6 (0.051)		
$CH_2{=}CH_2$	CH_4 (0.301),	C_2H_6 (0.045),	C_3H_8 (0.011)
cyclo-C_3H_6	C_2H_6 (0.272),	C_3H_8 (0.090),	CH_4 (0.029)

TiO_2: 0.30 g, reactant: 63.0 μmol.

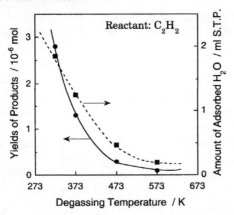

Figure 5.15. Effect of the degassing temperature of TiO_2 upon the yields of photoreaction of CH≡CH at 300 K

molecular water and chemisorbed hydroxyl groups, the former being completely removed by evacuation up to 473 K. These results, therefore, clearly suggest that water molecules, and not hydroxyl groups, are responsible for the photocatalytic reaction of TiO_2 catalysts. Similar photohydrogenation experiments have been carried out by using D_2O in place of H_2O. The D content in methane formed by photohydrogenation is also evidence that the non-dissociated adsorbed water is responsible for the photocatalytic hydrogenation reactions [96, 97].

UV irradiation of TiO_2 brings about electron–hole pairs which can be represented as a localised electron (Ti^{3+}) and hole (O_l^- and/or ·OH radical). Some portion of these species are separated from each other, being trapped by H^+, which will be supplied from more acidic surface hydroxyl groups, and OH^- ions to form H and ·OH radicals, respectively. The interaction of alkene (or alkyne) with a pair of these active species may result in hydrogenation accompanied by the fission of the C–C bond and the formation of oxygen-containing compounds. The rate of the photohydrogenolysis of alkenes increased with decreasing the ionisation potential of the alkene, suggesting that the interaction of alkene with the correlated Ti^{3+} (or H atom) and ·OH radical (or O_l^-) pairs involves an electron transfer from alkene to the ·OH moiety, i.e., the positive trapping holes of the active sites.

3.2 PHOTOCATALYTIC HYDROGENATION ON Pt/TiO$_2$

It has been established that the activity of semiconducting photocatalysts is markedly enhanced by the addition of small amounts of metals such as Pt on the catalyst [98–101]. Such an enhancement in the activity can be explained by concept that the electrons which are generated in semiconductors by photons easily transfer to the loaded metal particles, resulting in a decrease in electron–hole recombination. Although many studies deal with this concept, there has been little direct evidence to clarify the metal-loading effects on the primary processes of the gas-phase photocatalytic reactions [102–104].

Table 5.6 shows the results of the photocatalytic reaction of $CH_3C \equiv CH$ with H_2O on Pt-loaded TiO_2 (Pt/TiO$_2$) and unloaded TiO_2. With the TiO_2 catalyst, the major products arise from hydrogenation accompanied by the fission of the C–C bond (photohydrogenolysis). On the other hand, with the Pt/TiO$_2$ catalyst, the yields of the products from hydrogenation without the C–C bond fission (photo-hydrogenation) increase markedly, its extent being strongly dependent on the amount of Pt-loaded and also on the particle size of the TiO_2 photocatalysts. Similar results were obtained with other unsaturated hydrocarbons such as $CH_2=CH_2$, $CH_3CH=CH_2$, $CH \equiv CH$, etc. Table 5.6 also shows that there was little or no effect of Pt-loading on the photocatalytic reaction on the Pt/TiO$_2$ catalysts which were reduced with H_2 at 773 K, i.e., the SMSI behaviour is observed only in the photocatalytic reaction on the reduced Pt/TiO$_2$ catalysts at 473 K.

As shown in Figure 5.16, with the TiO_2 catalyst, the ESR signal attributed to the Ti^{3+} ions appears, its intensity increasing linearly with the UV irradiation time. With the Pt/TiO$_2$ catalysts, little or no change in the ESR signal due to the Ti^{3+} ions can be observed [102]. These results suggest that with Pt/TiO$_2$ catalysts photo-produced electrons easily transfer from TiO_2 to the Pt moieties. It can be proposed that an enhancement of the reaction, $H^+ + e^- \rightarrow H$, by the catalytic effect of Pt

Table 5.6. Products distribution and their yields in the photocatalytic reaction of $CH_3C \equiv CH$ with H_2O at 295 K

Catalyst type[a]	Pt content (wt%)	Products (10^{-9} mol/m^2 h)					
		CH$_4$	C$_2$H$_6$	C$_2$H$_4$	C$_3$H$_6$	C$_3$H$_8$	C$_3$/C$_2$[b]
TIO-4[c]	0	1.90	12.3	0.40	1.10	0.50	0.11
TIO-4[c]	0.4	2.10	12.7	0.54	4.04	0.35	0.29
TIO-4[c]	1.0	2.20	13.2	0.52	14.2	0.38	0.92
TIO-4[d]	1.0	2.20	12.2	0.50	1.15	0.48	0.11
TIO-4[c]	4.0	2.40	12.7	0.50	15.7	0.40	1.03
TIO-4[d]	4.0	1.90	12.2	0.39	1.20	0.47	0.12
Syn-55 Å[c]	0	6.02	29.0	—	—	—	0.004
Syn-55 Å[c]	4.0	2.89	16.0	—	154.0	—	9.6
Syn-120 Å[c]	0	0.91	4.2	—	0.02	—	0.004
Syn-120 Å[c]	4.0	0.06	0.6	—	13.5	—	25.0
Syn-400 Å[c]	0	0.68	2.9	—	—	—	0.003
Syn-400 Å[c]	4.0	trace	0.1	—	5.6	—	56.0

[a] TIO-4: standard TiO_2 catalyst (JRC-TIO-4) supplied from Japanese Catalysis Society, Syn: Anatase TiO_2 powders prepared by a precipitation method.
[b] $(C_3/C_2) = (C_3H_6 + C_3H_8)/(CH_4 + C_2H_6 + C_2H_4)$
[c] H_2 reduction was carried out at 473 K.
[d] H_2 reduction was carried out at 773 K.

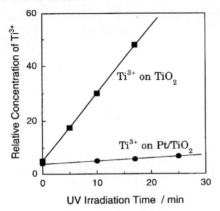

Figure 5.16. Growth of the ESR signal of the photoformed Ti^{3+} ions over Pt-loaded and unloaded TiO_2 at 77 K

brings about such electron transfers. A photoelectrochemical process is predominant, where the reduction of H^+ on the Pt particle and the oxidation by $\cdot OH$ on TiO_2 brings about photohydrogenation and the formation of oxidation products, respectively.

As shown in Figure 5.17, there is little or no change in the rate of photocatalytic reactions of the various hydrocarbons on the Pt/TiO_2 catalysts while the corresponding values for unloaded TiO_2 catalysts increase with the lowering of the ionisation potential of the hydrocarbons. These results suggest that the interaction between alkenes and photo-generated (electron–hole) pairs in TiO_2 catalysts plays an important role in photohydrogenolysis, its extent being stronger with hydrocarbons having lower ionisation potential. On the other hand, the photohydrogenation required no such interaction, resulting in similar reaction rates throughout all the alkenes. Thus, the following primary processes can be proposed for photocatalytic reactions on the Pt/TiO_2 and TiO_2 catalysts, respectively [102].

Unloaded TiO_2 : Pt/TiO_2

$e^- + H^+ \longrightarrow H$ $e^- + H^+ \rightarrow H\cdot$ (on Pt)

close existence

$h^+ + OH^- \longrightarrow OH$ $h^+ + OH^- \rightarrow \cdot OH$ (on TiO_2)

The studies of Fujishima *et al.* on the photocatalytic hydrogenation of ethylene on various bimetal-loaded TiO_2 powders and on monometal loaded samples, using the metal elements of Pd, Pt, Cu and Ni must also be considered [105]. For the selective production of ethane, a combination of Pt and Cu as co-catalysts showed their collective effect in the hydrogenation reaction. When the selectivity of produced C_2H_6 to CH_4 was improved by the optimum ratio of Pt to Cu on TiO_2, the

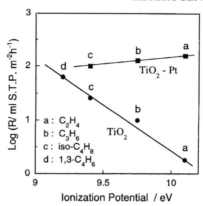

Figure 5.17. Relationship between the ionisation potential of reactants and the rates of their photocatalytic reactions on Pt/TiO$_2$ and unloaded TiO$_2$ at 295. (a: CH$_2$=CH$_2$, b: CH$_3$CH=CH$_2$, c: i-C$_4$H$_8$, d: 1,3-C$_4$H$_6$.)

volume of the concurrently evolved H$_2$ was diminished. This suggests that the Pt surface which was spatially controlled with the codeposited Cu afforded a desirable geometry for the photohydrogenation of C$_2$H$_4$. Furthermore, the additive metals on the catalyst surface prevented incident light from yielding a high surface density of O$^-$ or ·OH. Thus, the fission of the C=C bond on Cu/Pt/TiO$_2$ can be expected to be minimised under moderate illumination.

3.3   SIZE QUANTISATION EFFECT

Small colloidal particles such as CdS and ZnS exhibit large blue shifts in the absorption and photoluminescence spectra, indicating that the physical and chemical properties of small semiconductor particles are different from those of bulk materials [106–108]. The electronic properties of small semiconductors can be said to be dependent upon the crystallite size and shape due to the quantised motion of the electron and hole in a confined shape. The photocatalytic activity of the extremely small particles of TiO$_2$ was investigated as a function of their particle size.

As shown in Figure 5.18, it can also be seen that the quantum yields on anatase type TiO$_2$ for the photohydrogenolysis of CH$_3$C≡CH with water increase when the particle size is decreased and a significant increase in yields can be observed for both rutile and anatase type TiO$_2$ catalysts [39]. It is clear that these features correspond to those in the blue shifts of the absorption band when compared to those of the bulk TiO$_2$ crystallites rather than in the BET surface area.

UV irradiation of the small TiO$_2$ catalyst in the presence of CH$_3$C≡CH and H$_2$O at 77 K was found to lead to the appearance of ESR spectra attributed to two

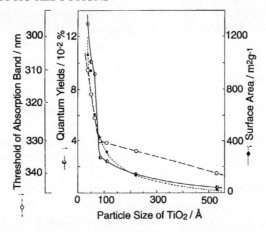

Figure 5.18. Effects of the particle size of TiO_2 (anatase) catalysts upon the quantum yields of the photocatalytic hydrogenolysis of $CH_3C{=}CH$ with water, threshold of absorption band, and BET surface area of the catalysts

different unstable carbon-centred radicals ($CH{\equiv}{\cdot}CHads$, ${\cdot}CHads$) as well as simultaneously photoformed O^- ($g = 2.0014$) and Ti^{3+} ($g_\perp = 1.988$). With the small particle TiO_2, which exhibits higher photocatalytic activity, the concentrations of photoformed carbon-centred radicals and of the Ti^{3+} ions were found to be high [39]. An increase in the photocatalytic activity of TiO_2, with a decrease in the particle size, especially in the range of less than 100 Å diameter, was found to be closely associated with the size quantisation effect.

The size quantisation effect results in the electronic modification in the TiO_2 particle and leads to an enhancement in the activity of both the photo-formed electron and hole species and/or a suppression of the radiative transfer of the absorbed photon energy. In other words, the existence of extremely small particles in which the photo-formed electron (Ti^{3+}) and hole (${\cdot}OH$ and/or O^-) are close to each other causes their interaction to be significant. This situation results in a high efficiency in the photocatalytic reactions, as well as a well-balanced contribution of photo-produced electrons and holes to the reductive and oxidative surface reactions, respectively.

3.4 ANCHORED TITANIUM OXIDE CATALYSTS

Highly dispersed titanium oxide anchored onto Vycor glass have been prepared using a CVD method to investigate their photocatalytic activity [58,59,109]. UV irradiation of the anchored catalyst in the presence of alkyne and water led to the formation of hydrogenation products accompanied by the fission of the carbon–carbon bond (hydrogenolysis).

Figure 5.19 shows that the yields of the photohydrogenolysis of $CH_3C\equiv CH$ with H_2O vary in a manner similar to the intensity of the photoluminescence of the catalyst [59]. Since the photoluminescence originates from the recombination of the correlated electron and hole which constitute the excited complex of the $(Ti^{3+}–O^-)$ pair, this complex was found to play a significant role in the photohydrogenolysis being in good agreement with the results obtained with the small particle TiO_2 catalyst. The photocatalytic activity of the anchored titanium oxide is greater than that of bulk TiO_2 by 2–3 orders of magnitude. Furthermore, the selectivity to form products with the fission of the carbon–carbon bond (hydrogenolysis) against the products for hydrogenation on the anchored catalysts was higher than the bulk TiO_2 catalyst.

An ESR technique was used to investigate the local structure of the titanium oxide species anchored onto support surfaces by monitoring the Ti^{3+} ions which were formed by the photoreduction of the oxide with H_2 at 77 K (4) [58, 60].

$$Ti^{4+} - O^{2-} + H_2 \rightarrow Ti^{3+} - OH^- + H \tag{5.4}$$

As shown in Figure 5.20, the ESR spectrum of the anchored titanium oxide species is composed of two different types of Ti^{3+}, one with a g-value of $g_\perp = 1.9790$ and the other 1.9640. However, the addition of H_2O onto the sample led to a decrease in the intensity of the ESR signal at $g_\perp = 1.9790$, the extent depending on the amount of H_2O added. Excess amounts of H_2O led to the complete disappearance of this signal. Simultaneously, the intensity of the signal at $g_\perp = 1.9640$ increased with the amounts of H_2O added. It was found that the amount of Ti^{3+} which disappeared

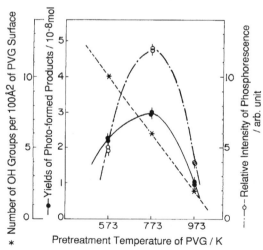

Figure 5.19. Effect of the pretreatment temperature of Vycor glass upon the yields of photodehydrogenation reaction of $CH_3C\equiv CH$ with water and of photoluminescence of the titanium oxide anchored onto Vycor glass at 300 K

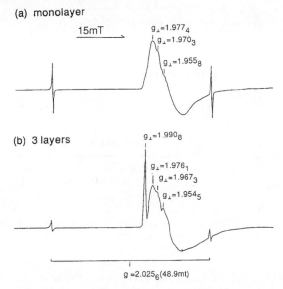

(a) monolayer

15mT →

$g_\perp = 1.977_4$
$g_\perp = 1.970_3$
$g_\perp = 1.955_8$

(b) 3 layers

$g_\perp = 1.990_8$
$g_\perp = 1.976_1$
$g_\perp = 1.967_3$
$g_\perp = 1.954_5$

$g = 2.025_6 (48.9mt)$

Figure 5.20. ESR spectra (at 77 K) of photo-reduced Ti^{3+} ions on the titanium oxide anchored onto Vycor glass with different number of (Ti-O) layers

was roughly equal to that of the Ti^{3+} ions which appeared after the addition of H_2O and these changes were reversible. These results clearly indicate that Ti^{4+} ions, i.e., the titanium oxide species are present in a tetrahedral coordination [58, 60]. The EXAFS studies at the Ti K-edge also indicate the presence of isolated titanium oxide species in a tetrahedral coordination [63].

We have, therefore, concluded that the high photocatalytic activity for photocatalytic hydrogenolysis of alkyne with water on the anchored titanium oxides are associated with the presence of well-dispersed homogeneous Ti^{4+} ions and/or their coordinative unsaturation, which result in less efficient radiationless deactivation of absorbed photon energy and/or in a peculiar coordination capacity of the anchored Ti^{4+} ions.

3.5 Ti/Si AND Ti/Al BINARY OXIDE CATALYSTS

For the hydrogenolysis and hydrogenation of alkynes and alkenes, the photocatalytic activity of the titanium oxide species in the Ti/Si and Ti/Al binary oxides prepared by a coprecipitate method was found to be enhanced in the regions of lower Ti content [110–112]. As shown in Figure 5.21, X-ray, photoluminescence, and XPS measurements of the surface of the catalysts indicated that Ti ions are enriched on the surface of the catalysts having a lower Ti content and present separately from each other as small titanium oxide moieties in the SiO_2 or in the

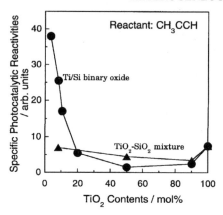

Figure 5.21. Effects of TiO_2 contents on the specific photocatalytic activities of Ti/Si binary oxide catalysts and mixtures of TiO_2 and SiO_2

Al_2O_3 matrix. The EXAFS and ESR measurements have also revealed that these species are isolated titanium oxide moieties having a tetrahedral coordination. This was found to result in an enhancement of the photocatalytic activity of the titanium oxide species and due to the diminished radiationless transfer of the photon energy absorbed by the titanium oxide moiety.

4 CONCLUSIONS

The characteristic features of the photocatalytic reduction of CO_2 with H_2O on various types of active titanium oxide catalysts and the photocatalytic hydrogenation and/or hydrogenolysis of alkynes and alkenes with water on various types of titanium oxide catalysts have been clarified in this chapter.

UV irradiation of active titanium oxide catalysts in the presence of CO_2 and H_2O at 275 K led to the photocatalytic reduction of CO_2. The reactions on TiO_2 powders and Ti/Si binary oxides prepared by the sol–gel method produced methane as the major product while, on the highly dispersed titanium oxide anchored on porous glass and zeolites, the formations of CH_4, CH_3OH, C_2-compounds, and CO were observed as the major products. The yields of the photocatalytic reactions strongly depended on the type of catalyst used, the value of CO_2/H_2O, and the reaction temperature. *In situ* spectroscopic studies of the system indicated that the photocatalytic reduction of CO_2 with H_2O is linked to a much higher reactivity of the charge transfer excited state, i.e. $(Ti^{3+}\text{-}O^-)^{3*}$ of the tetrahedral coordinated titanium oxides species formed on the surface. Based on the detection of the reaction intermediate species such as Ti^{3+}, H· atoms, and C· radicals, the reaction mechanism for the photocatalytic reduction of CO_2 with H_2O has been proposed on a molecular scale.

Photocatalytic reactions of alkynes and alkenes with water have been investigated over TiO_2 and Pt-loaded TiO_2 catalysts. With TiO_2, the major products were formed by photohydrogenolysis. The interaction of the reactants with the photoformed electron (Ti^{3+}) and hole ($\cdot OH$) pair state plays a significant role in the hydrogenolysis reaction. With Pt-loaded TiO_2 catalysts, a significant enhancement of photohydrogenation could be observed. A photoelectrochemical process was predominant where the reduction of H^+ on the Pt particle and oxidation of OH^- on TiO_2 brings about the photohydrogenation without the bond fission and the formation of oxidation products, respectively. The photocatalytic activity of small particle TiO_2 catalysts increased as the diameter of the TiO_2 particle became smaller, especially below 100 Å. The size quantisation effect was in effect in these extremely small particle TiO_2, resulting in the electronic modification of TiO_2 which led to an enhancement of the activity of the local electrons (Ti^{3+}) and holes ($\cdot OH$ or O_l^-). The higher photocatalytic activity of the anchored titanium oxide catalysts and the titanium oxide in the Ti/Si and the Ti/Al binary oxides in the regions of lower Ti content was closely linked to the presence of well-dispersed homogeneous Ti^{4+} ions and/or their coordinative unsaturation.

These results clearly demonstrate that the unsaturated surface Ti^{4+} ions located in specific surface sites play a significant role in the photocatalytic reactions on TiO_2 catalysts. Ions of high coordination have a larger number of bonds to the oxide and couple more strongly with the phonon transitions of the lattice, providing a high probability of non-radiative decay. On the other hand, on the lower coordinative surface ions, non-radiative decay is less efficient and the radiative decay pathways are more predominant, if there are no reactant molecules. As a result, higher photocatalytic activity as well as photoluminescence are achieved on the titanium oxide species having lower coordinated surface ions.

REFERENCES

[1] M. Anpo and H. Yamashita, in *Surface Photochemistry*, ed. by M. Anpo (John Wiley & Sons, 1996) p. 117.

[2] *Energy Resources through Photochemistry and Catalysis* ed. by M. Grätzel (Academic Press, New York, 1983).

[3] M. Calvin, *Acc. Chem. Res.*, **11**, 369 (1978).

[4] G. M. Woodwell, J. E. Hobbie, R. A. Houghton, J. M. Melillo, B. Moore, B. J. Peterson, and F. R. Shaver, *Science*, **222**, 1081 (1983).

[5] D. E. Gushee, *Chemtech*, 470 (1989).

[6] J. Kondo, T. Inui, and K. Wasa, *Proceedings of the Second International Conference on Carbon Dioxide Removal* (Pergamon, London, 1995).

[7] J. C. Hemminger, R. Carr, and G. A. Somorjai, *Chem. Phys. Lett.*, **57**, 100 (1978).

[8] J. C. Hemminger, R. Carr, W. J. Low, and G. A. Somorjai, *Adv. Chem. Ser.*, **184**, 233 (1980).

[9]   T. Inoue, A. Fujishima, S. Konishi, and K. Honda, *Nature*, **277**, 637 (1979).
[10] B. Aurian-Blajeni, M. Halmann, and J. Manassen, *Solar Energy*, **25**, 165 (1980).
[11] M. Halmann, in *Energy Resources through Photochemistry and Catalysis*, ed. by M. Grätzel (Academic Press, New York, 1983), p. 507.
[12] M. Halmann, M. Ulman, and B. A.-Blajeni, *Solar Energy*, **31**, 429 (1983).
[13] M. Halmann, V. Katzir, E. Borgavello, and J. Kiwi, *Solar Energ. Mater.*, **10**, 85 (1984).
[14] M. Ulmann, B. Aurian-Blajeni, and M. Halmann, *Chemtech.*, 235 (1984).
[15] M. Ulmann, A. H. A. Tinnemens, A. Mackor, B. Aurian-Blajeni, and M. Halmann, *Int. J. Solar Energy*, **1**, 213 (1982).
[16] M. Aresta, E. Quaranta, and I. Tommasi, in *Photochemical Conversion and Storage of Solar Energy*, ed. by E. Pelizzetti and M. Schiavello (Kluwer Academic Publishers, The Netherlands, 1991), p. 577.
[17] K. Chandrasekaran and J. K. Thomas, *Chem. Phys. Lett.*, 99 (1983).
[18] K. Tannakone, *Solar Energ. Mater.*, **10**, 85 (1984).
[19] M. W. Raphael and M. A. Malati, *J. Chem. Soc., Chem. Commun.*, 1418 (1987).
[20] P. Albers and J. Kiwi, *New J. Chem.*, **14**, 135 (1990).
[21] Z. Goren, I. Willne, A. J. Nelson, and A. J. Frank, *J. Phys. Chem.*, **94**, 3738 (1990).
[22] F. Saladin, L. Forss, and I. Kamber, *J. Chem. Soc., Chem. Commun.*, 533 (1995).
[23] R. K. Thampi, J. Kiwi, and M. Grätzel, *Nature*, **237**, 506 (1987).
[24] N. Vlachopoulos, P. Liska, J. Augustynski, and M. Grätzel, *J. Am. Chem. Soc.*, **110**, 1216 (1988).
[25] M. Anpo and K. Chiba, *J. Mol. Catal.*, **74**, 207 (1992).
[26] M. Anpo, H. Yamashita, S. Kawasaki, and Y. Ichihashi, *Sekiyu Gakkaishi*, **38**, 300 (1995).
[27] M. Anpo, H. Yamashita, Y. Ichihashi, and S. Ehara, *J. Electroanal. Chem.*, **396**, 21 (1996).
[28] M. Anpo, *Solar Energ. Mater. Solar Cells*, **38**, 221 (1995).
[29] H. Yamashita, H. Nishiguchi, N. Kamada, M. Anpo, Y. Teraoka, H. Hatano, S. Ehara, K. Kikui, L. Palmisano, A. Sclafani, M. Schiavello and M. A. Fox, *Res. Chem. Intermed.*, **20**, 815 (1994).
[30] H. Yamashita, N. Kamada, H. He, K. Tanaka, S. Ehara, and M. Anpo, *Chem. Lett.*, 855 (1994).
[31] H. Yamashita, A. Shiga, S. Kawasaki, Y. Ichihashi, S. Ehara, and M. Anpo, *Energy Conv. Manag.*, **36**, 617 (1995).
[32] S. Yamamura, H. Kojima, J. Iyoda, and W. Kawai, *J. Electroanal. Chem.*, **225**, 287 (1987).
[33] S. Yamamura, H. Kojima, J. Iyoda, and W. Kawai, *J. Electroanal. Chem.*, **247**, 333 (1988).
[34] K. Ogura, M. Kanemoto, J. Yano, and Y. Sakata, *J. Photochem. Photobiol. A: Chem.*, **66**, 91 (1991).
[35] K. Hirano, K. Inoue, and T. Yasu, *J. Photochem. Photobiol. A: Chem.*, **64**, 255 (1992).
[36] O. Ishitani, C. Inoue, Y. Suzuki, T. Ibusuki, *J. Photochem. Photobiol, A: Chem.*, **72**, 269 (1993).
[37] K. Sayama and H. Arakawa, *J. Phys. Chem.*, **97**, 531 (1993).
[38] M. Watanabe, *J. Jpn Surf. Sci.*, **13**, 365 (1992).
[39] M. Anpo, T. Shima, S. Kodama, and Y. Kubokawa, *J. Phys. Chem.*, **91**, 4305 (1987).
[40] M. Anpo, *Res. Chem. Intermed.*, **11**, 67 (1989).
[41] M. Anpo and Y. Kubokawa, *Res. Chem. Intermed.*, **8**, 105 (1989).
[42] M. Anpo, *Surface Photochemistry* (John Wiley & Sons, London, 1996).

[43] *Photochemistry on Solid Surfaces* ed. by M. Anpo and T. Matsuura, (Elsevier, Amsterdam, 1989).

[44] *Hikari-Shokubai (Photocatalysis)* ed. by Y. Kubokawa, K. Honda, and Y. Saito, (Asakura-shoten, Tokyo, 1988).

[45] *Photocatalysis* ed. by N. Serpone and E. Pelizzetti, (John Wiley & Sons, New York, 1988).

[46] *Photochemical Conversion and Storage of Solar Energy* ed. by E. Pelizzetti and M. Schiavello, (Kluwer, Dordrecht, 1991).

[47] A. L. Linsebigler, G. Lu, and J. T. Yates, *Chem. Rev.*, **95**, 735 (1995).

[48] M. Anpo, M. Tomonari and M. A. Fox, *J. Phys. Chem.*, **93**, 7300 (1989).

[49] K. W. Frese, *J. Electrochem. Soc.*, **138**, 3338 (1991).

[50] J. Raskö and F. Solymosi, *J. Phys. Chem.*, **98**, 7147 (1994).

[51] F. Solymosi, *J. Mol. Catal.*, **65**, 337 (1991).

[52] F. Solymosi and G. Klivényi, *Catal. Lett.*, **22**, 337 (1993).

[53] G. Heiland and H. Lueth, in *The Physics of Solid Surfaces and Heterogeneous Catalysis III*, ed. by D. A. King and D. P. Woodruff (Elsevier, Amsterdam, 1984) pp. 137.

[54] G. A. Somorjai and F. Zaera, *J. Phys. Chem.*, **86**, 3070 (1982).

[55] J. C. S. Wong, A. L. Linsebigler, G. Lu, J. F. Fan, and J. T. Yates, *J. Phys. Chem.*, **99**, 335 (1995).

[56] T. Yamada, T. Misono, K. Tanaka, and Y. Murata, *J. Vac. Sci. Technol.*, **A7**, 2308 (1989).

[57] P. A. Dilara and J. M. Vohs, *J. Phys. Chem.*, **97**, 12919 (1993).

[58] M. Anpo, N. Aikawa, Y. Kubokawa, M. Che, C. Louis, and E. Giamello, *J. Phys. Chem.*, **89**, 5689 (1985).

[59] M. Anpo, N. Aikawa, Y. Kubokawa, M. Che, C. Louis, and E. Giamello, *J. Phys. Chem.*, **89**, 5017, (1985).

[60] M. Anpo, T. Shima, T. Fujii, and M. Che, *Chem. Lett.*, 65 (1987).

[61] M. Anpo, T. Shima, and M. Che, *Chem. Express*, **3**, 403 (1988).

[62] H. Yamashita, Y. Ichihashi, and M. Anpo, *J. Surf. Sci. Soc. Jpn.*, **16**, 194 (1995).

[63] H. Yamashita, Y. Ichihashi, M. Harada, G. Stewart, M. A. Fox, and M. Anpo, *J. Catal.*, **158**, 97 (1996).

[64] H. Yamashita, Y. Ichihashi, C. Louis, M. Che, and M. Anpo, *J. Phys. Chem.*, **100**, 16041 (1996).

[65] H. Yamashita, Y. Ichihashi, S. C. Zhang, T. Tatsumi, Y. Matsumura, Y. Souma, and T. Tatsumi, *Appl. Surf. Sci.*, in press.

[66] Y. Ichihashi, H. Yamashita, and M. Anpo, *J. Phys IV (Colloques)*, in press.

[67] H. Yamashita, S. Kawasaki, and M. Anpo, *Shokubai (Catalyst)* **36**, 440 (1994).

[68] A. Henglein, M. Gutierrez, *Ber. Bunsen-Ges. Phys. Chem.*, **87**, 852 (1983).

[69] B. R. Eggins, J. T. S. Irvin, E. P. Murphy, and J. Grimshaw, *J. Chem. Soc., Chem. Commun.*, 1123 (1988).

[70] J. M. Nedeljkovc, M. T. Nenadovi, D. I. Micic, and A. J. Nozik, *J. Phys. Chem.*, **90**, 12 (1986).

[71] D. W. Bahnemann, C. Kormann, and M. R. Hoffmann, *J. Phys. Chem.*, **90**, 12 (1986).

[72] L. Spanhel, M. Haase, H. Weller, and A. Henglein, *J. Am. Chem. Soc.*, **109**, 5649 (1987).

[73] T. Shiragami, C. Pac, and S. Yanagida, *J. Phys. Chem.*, **94**, 504 (1990).

[74] H. Yoneyama, S. Haga, and S. Yamanaka, *J. Phys. Chem.*, **93**, 4833 (1989).

[75] A. Henglein, M. Gutierrez, and M. Fischer, *Ber. Bunsen-Ges. Phys. Chem.*, **88**, 170 (1984).

[76] H. Inoue, T. Torimoto, T. Sakata, H. Mori, and H. Yoneyama, *Chem. Lett.*, 1483 (1990).
[77] H. Inoue, R. Nakamura, and H. Yoneyama, *Chem. Lett.*, 1227 (1994).
[78] H. Inoue, H. Moriwaki, K. Maeda, and H. Yoneyama, *J. Photochem. Photobiol. A: Chem.*, **86**, 191 (1995).
[79] S. Kuwabata, K. Nishida, R. Tsuda, H. Inoue, and H. Yoneyama, *J. Electrochem.*, **141**, 1498 (1994).
[80] S. Yanagida, M. Yoshida, T. Shiragami, C. Pac, H. Mori, and H. Fujita, *J. Phys. Chem.*, **94**, 3104 (1990).
[81] M. Kanemoto, T. Shiragami, C. Pac, and S. Yanagida, *Chem. Lett.*, 931 (1990).
[82] M. Kanemoto, T. Shiragami, C. Pac, and S. Yanagida, *J. Phys. Chem.*, **96**, 3521 (1992).
[83] Y. Wada and S. Yanagida, *J. Surf. Sci. Soc. Jpn.*, **16**, 180 (1995).
[84] S. Matsuoka, H. Fujii, C. Pac, and S. Yanagida, *Chem. Lett.*, 1501 (1990).
[85] S. Matsuoka, H. Fujii, T. Yamada, C. Pac, A. Ishida, S. Takamuku, M. Kusaba, N. Nakashima, S. Yanagida, K. Hashimoto, and T. Sakata, *J. Phys. Chem.*, **95**, 5802 (1991).
[86] S. Matsuoka, T. Kohzuki, C. Pac, and S. Yanagida, *Chem. Lett.*, 2047 (1990).
[87] S. Matsuoka, T. Kohzuki, C. Pac, A. Ishida, S. Takamuku, M. Kusaba, N. Nakashima, and S. Yanagida, *J. Phys. Chem.*, **96**, 4437 (1992).
[88] S. Matsuoka, K. Yamamoto, C. Pac, and S. Yanagida, *Chem. Lett.*, 2099 (1991).
[89] S. Matsuoka, K. Yamamoto, T. Ogata, M. Kusaba, N. Nakashima, E. Fujita, and S. Yanagida, *J. Am. Chem. Soc.*, **115**, 601 (1993).
[90] T. Ogata, S. Yanagida, B. S. Brunschwig, and E. Fujita, *J. Am. Chem. Soc.*, **117**, 6708 (1995).
[91] T. Ogata, Y. Yamamoto, Y. Wada, K. Murakoshi, M. Kusada, N. Nakashima, A. Ishida, S. Takamuku, and S. Yanagida, *J. Phys. Chem.*, **99**, 11916 (1995).
[92] A. H. Boonstra and C. A. H. Mutsaers, *J. Phys. Chem.*, **79**, 2025 (1975).
[93] G. N. Schrauzer and T. D. Guth, *J. Am. Chem. Soc.*, **99**, 7189 (1977).
[94] S. Sato and J. M. White, *Chem. Phys. Lett.*, **70**, 131 (1980).
[95] M. Anpo and Y. Kubokawa, *J. Catal.*, **75**, 204 (1982).
[96] C. Yun, M. Anpo, S. Kodama, and Y. Kubokawa, *J. Chem. Soc., Chem. Commun.*, 609 (1980).
[97] M. Anpo, N. Aikawa, S. Kodama, and Y. Kubokawa, *J. Phys. Chem.*, **88**, 2569 (1984).
[98] B. Kraeutler and A. J. Bard, *J. Am. Chem. Soc.*, **99**, 7729 (1979).
[99] H. Reiche and A. J. Bard, *J. Am. Chem. Soc.*, **101**, 312 (1979).
[100] F. T. Wagner and G. A. Somorjai, *J. Am. Chem. Soc.*, **102**, 5494 (1988).
[101] H. V. Damme and W. K. Hall, *J. Am. Chem. Soc.*, **101**, 4373 (1979).
[102] M. Anpo, N. Aikawa, and Y. Kubokawa, *J. Phys. Chem.*, **88**, 3998 (1984).
[103] M. Anpo and M. Tomonari, *Denki Kagaku*, **57**, 1219 (1989).
[104] M. Anpo, K. Chiba, M. Tomonari, S. Coluccia, M. Che, and M. A. Fox, *Bull. Chem. Soc. Jpn.*, **64**, 543 (1991).
[105] R. Baba, S. Nakabayashi, A. Fujisjima, and K. Honda, *J. Am. Chem. Soc.*, **109**, 2273 (1987).
[106] L. E. Brus, *J. Phys. Chem.*, **90**, 2555 (1986).
[107] F. Fojtik, U. Koch, and A. Henglein, *Ber Bunsen-Ges. Phys. Chem.*, **88**, 170 (1984).
[108] A. J. Nozik, F. Williams, M. T. Nenadovic, T. Rajh, and O. I. Micic, *J. Phys. Chem.*, **89**, 397 (1985).
[109] M. Anpo, N. Aikawa, and Y. Kubokawa, *J. Chem. Soc., Chem. Commun.*, 644 (1984).

[110] M. Anpo, H. Nakaya, S. Kodama, Y. Kubokawa, K. Domen, and T. Onishi, *J. Phys. Chem.*, **90**, 1633 (1986).
[111] S. Kodama, H. Nakaya, M. Anpo, and Y. Kubokawa, *Bull. Chem. Soc. Jpn.*, **58**, 3645 (1985).
[112] M. Anpo, H. Nakaya, S. Kodama, Y. Kubokawa, and K. Domen, *J. Phys. Chem.*, **92**, 438 (1988).

6 Heterogeneous Photocatalytic Reactors: An Assessment of Fundamental Engineering Aspects

V. AUGUGLIARO, V. LODDO and M. SCHIAVELLO

Dipartimento di Ingegneria Chimica dei Processi e dei Materiali Università di Palermo, Viale delle Scienze, 90128 Palermo, Italy

1 INTRODUCTION

Heterogeneous photocatalytic processes are receiving a great interest in the pollution control area owing to their ability for eliminating contaminant species present in aqueous and in gaseous effluents [1–3]. The main advantages of this method are its non-specificity and the possibility of treating effluents with very low concentrations of contaminants without which the reaction rate decreases to negligible values.

It is well known that the absorption of light by a great variety of chemical species occurs in the far-UV region of the radiation spectrum, this being the cause of their stability in the environment. In order to excite these chemicals with radiation of low energy (and cost), as near-UV radiation that is also present in the sunlight, the presence is necessary of a solid photocatalyst [4, 5] which is generally a semiconductor of suitable band-gap energy.

The utilisation of radiation and solid catalysts in a synergic effect for performing chemical transformations of species present in liquid or gaseous mixtures makes the heterogeneous photocatalytic process much more complex than the homogeneous photochemical one as another phase is added to the system. For example, when stirred dispersions are used, the presence of suspended solid particles determines not only fluidodynamic complications, since the solid affects momentum transfer, but also questions of opacity, scattering and depth of radiation penetration are to be taken into account.

The main components of a photocatalytic process are indeed the photoreactor and the radiation source [6]. For thermal and catalytic processes the reactors are generally chosen on the basis of the following parameters: (i) the mode of operation; (ii) the phases present in the reactor; (iii) the flow characteristics; (iv) the needs of heat exchange; (v) the composition and the operative conditions of the reacting mixture, which affect the selection of materials of construction. For selecting the type of heterogeneous photoreactor additional parameters must be considered owing to the fact that photons are needed for the occurrence of photoreaction. The selection of the construction material for the photoreactor must be generally done in order to allow the penetration of radiation into the reacting mixture. The choice of the radiation source is easier than that of the photoreactor. In principle, in order to improve the economics of the photocatalytic process, the radiation spectrum of the lamp used should coincide with the absorption spectrum of the reacting system. In the case of photocatalytic systems utilising semiconducting solids, the absorbed radiation energy should be equal to or higher than the band gap.

The main difficulties present in heterogeneous photocatalytic reactor engineering are linked to the fact that: (i) the fluidodynamics of a fluid–solid system is highly non-ideal so that modelling of the specific system is generally needed: and (ii) the kinetics of a heterogeneous photocatalytic reaction is affected in a complex way by the concentration and the radiation fields which are affected by the fluidodynamics. The coupling of models that describe the radiation emission, the volumetric radiation absorption and the fluidodynamics of the reacting system in the reactor determines that the design of a heterogeneous photoreactor is a very difficult task. Generally the set of conservation equations which model the photoreactor can be solved only numerically.

2   RADIATION SOURCES

The light source is a very important factor to be considered as the performance of a photoreactor is strongly dependent on the irradiation source. Different types of lamps allow generation of radiation with different ranges of wavelengths. As reported before, the choice of a particular lamp is made on the basis of the reaction energy requirements.

There are four main types of radiation sources. They are: (a) arc lamps; (b) fluorescent lamps; (c) incandescent lamps; (d) lasers.

In arc lamps the emission is obtained by a gas activated by collisions with electrons accelerated by an electric discharge. The activated gases are mercury and/ or xenon vapours.

In fluorescent lamps the emission is obtained by exciting an emitting fluorescent substance, deposited in the inner side of a cylinder, by an electric discharge occurring in the gas filling the lamps. Generally these lamps emit in the visible region, but the 'actinic' type ones have emission in the near-UV region. Of course the emission spectrum depends on the nature of the mixture of fluorescent substances used. Their power is quite small (up to 150 W).

In incandescent lamps, the emission is obtained by heating at very high temperature suitable filaments, of various nature, by current circulation.

The features of lasers are well known, especially for their frequent use in photochemistry and in many other fields. They produce coherent radiation of very high intensity.

In photocatalysis in general arc and fluorescent lamps are used for various reasons which were expressed in the preceding chapters.

For arc lamps and in particular for mercury lamps, a classification based on the pressure of Hg is made and it is as follows:

1. *Low-pressure Hg lamps.* This type of lamp contains Hg vapour at a pressure of about 0.1 Pa at 25 °C, emitting mainly at 253.7 and 184.9 nm.
2. *Medium-pressure Hg lamps.* This type of lamp has a radiation source containing mercury vapour at pressures ranging from 100 to several hundred kPa. Emission is mostly from 310 to 1000 nm with most intense lines at 313, 366, 436, 576 and 578 nm.
3. *High-pressure Hg lamps.* This type of lamp has a radiation source containing mercury at a pressure of 10 MPa or higher, which emits broad lines and a background continuous between about 200 and 1000 nm.
4. *Xenon and Hg-Xenon lamps.* In this type of lamp an intense source of ultraviolet, visible and near-IR radiation in a mixture of Hg and Xe vapours under high pressure is obtained. Xenon lamps are used to simulate the solar irradiating spectrum and the presence of Hg vapour increases the radiation intensity in the UV region.

Typical emission spectra are shown in Figures 6.1 and 6.2.

These lamps are usually cylindrical, with arc length increasing as pressure decreases and power increases. Power ranges from a few watts to about 60 000 W. Generally medium- and higher-pressure mercury lamps need to be cooled by circulating cooling liquids around them. Filtering solutions and glass filters can be

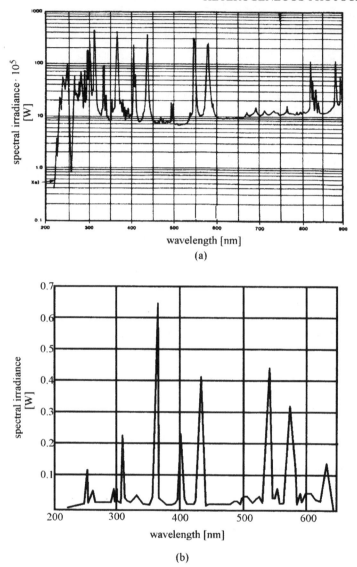

Figure 6.1. Emission spectra for higher and medium pressure arc lamp: (a) 1000 W high-pressure Hg–Xe lamp; (b) 125 W medium-pressure Hg lamp

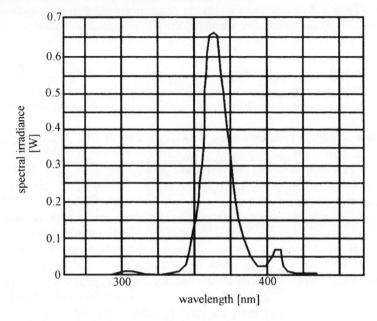

Figure 6.2. Emission spectrum of a 40 W actinic lamp

used with the aim of absorbing the wavelengths that should not reach the reacting medium.

3 FEATURES OF HETEROGENEOUS PHOTOREACTORS

Three types of ideal reactors are assumed in chemical engineering: the batch reactor, the plug flow reactor (PFR) and the continuous-flow stirred tank reactor (CSTR). A batch reactor has no input or output of mass; the stirring of the reactor ensures that there are no temperature or concentration gradients within the system volume. The concentration and the temperature will change with time due to the reaction, but at any time they are everywhere uniform. The reaction rate is also uniform and it is considered everywhere equal to the average value.

A PFR is usually visualised as a long tubular reactor. For the tube to be a PFR, three conditions must be satisfied: (i) the axial velocity profile is flat; (ii) there is complete mixing across the tube so that all the reactor variables are a function of the axial dimension of the reactor (named z); and (iii) there is no mixing in the axial direction. PFRs have spatial variations in concentration and temperature. Such systems are called distributed and analysis of their steady-state performance requires solution of differential equations.

The CSTR is a flow reactor in which the contents are mechanically agitated. If mixing caused by the agitator is adequate, the entering feed will be quickly dispersed through the vessel and the composition and temperature at any point will approximate the average composition and temperature. Perfect mixers have no spatial distribution of compositions and temperatures. Such systems are called lumped. The steady-state performance of lumped systems is determined by algebraic equations rather than the differential equations needed for distributed systems such as PFR.

For treating high-volume chemicals, flow reactors are usually preferred to batch reactors. Flow reactors are operated continuously; that is, at steady-state with reactants continuously entering the vessel and with products continuously leaving.

For heterogeneous photocatalytic reactions the contact among reactants, photons and catalysts must be maximised. Mixing and flow characteristics of the photoreactor may greatly enhance these contacts. If a fixed bed reactor is used, the irradiated aliquot of catalyst is limited to a thin layer and a large reactor volume is required. For liquid–solid and gas–solid systems, continuously stirred tank photoreactors and fluidised bed photoreactors, respectively, are the most suitable ones for enhancing contact efficiency even if their operation is quite expensive. The rate of photocatalytic reaction is greatly affected by stirring speed. The rate enhancement is not due to elimination of mass transport resistances, as expected in classical catalytic systems, since such considerations do not apply for most heterogeneous photoprocesses, which are characterised by low reaction rate with respect to mass transport rate. The enhancement is determined by the fact that on increasing stirring rate, the frequency of exposure of the catalyst particles to irradiation increases. The catalyst particles continuously receive diffuse radiation of reduced intensity due to absorption by other catalyst particles. They are directly irradiated intermittently due to shielding effect of particles which randomly intercept direct irradiation. By increasing stirring, the frequency with which the catalyst particles may be directly irradiated increases and, eventually, the reaction rate is enhanced.

The material of construction of photoreactors is chosen on the basis of the following considerations. If the reacting mixture can be processed under 'open sky', i.e. in contact with the atmosphere, the photoreactor, batch or continuous, may be an open vessel of whatever material and the irradiation can be performed from the open side (Figure 6.3). This is the case of a decontamination process which involves a liquid phase; the operation with a reactor in contact with air in this case is beneficial as oxygen is needed for the photoreaction to occur.

When the reacting mixture must be processed in a closed ambient, the requirement of light transmission determines the type of material. This is the case when gases are processed or the photoreacting mixture can contaminate (or can be contaminated by) air. In order to allow the penetration of radiation into the photoreactor, the photoreactor wall between the lamp and the reacting mixture must be made of a material transparent to the radiation. If near-UV radiation is utilised by the photoreacting mixture, Pyrex glass can be used but if UV radiation is needed,

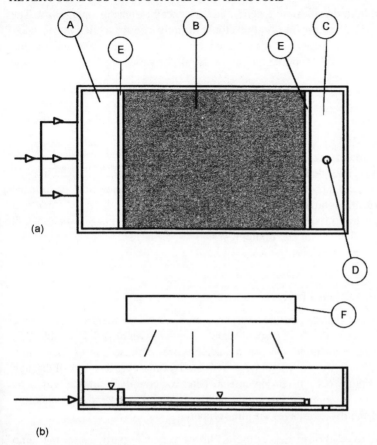

Figure 6.3. Fixed bed flat photoreactor: (a) upward view. A) feed zone; B) fixed catalytic bed; C) discharge zone; D) discharge hole; E) weirs; (b) lateral view. F) irradiation system

quartz glass must be used. It must be noted that, by increasing the thickness of photoreactor walls, the light transmission decreases so that the size of reactor and its operative conditions (temperature and pressure) are not independent variables. In order to utilise all the radiation emitted by the lamp, the transmission or absorption of radiation from photoreactor walls is avoided by wrapping them with reflecting surfaces such as aluminium foils.

3.1 PHOTOREACTOR GEOMETRIES

For photoreactors, the geometry and the spatial relation between reactor and light source are most important. The reason for that lies in the fact that geometry plays an important role in determining reaction yields as well as reactor operability. The geometrical configuration of a photoreactor is usually chosen to derive the maximum benefit from the pattern of irradiation taking into account shape and cooling requirements of available commercial lamps useful for wavelength required by the reaction.

The most usual photoreactor geometries are:

1. cylindrical;
2. parallel plate;
3. annular.

The irradiation may be normal or parallel to the reactor surface. In selecting the reactor geometry configuration it is necessary to determine the optical path of the light which will be obtained within the reactor. It is, in fact, the most important factor affecting the irradiation absorption by the reacting mixture and therefore it determines the efficiency of the photoprocess.

3.1.1   Immersion Well Photoreactor

This is the simplest type of photoreactor, used on a laboratory scale as well as on a pilot or production scale. It is a stirred tank reactor in which solid particles are generally suspended in liquid. One or more lamps are immersed in the suspension (in many cases conventional reactors are modified by mounting lamps well through their covers) (Figure 6.4). It can operate in batch or continuous mode way. The advantage of this geometrical configuration is simplicity and very high photonic efficiency, and the disadvantages are the following:

1. There is a possibility of depositing a film of very fine particles on the lamp surface so that the radiation of the dispersion decreases with time. To avoid this problem the cleaning of the lamp must be performed at fixed intervals of time by stopping the photoreactor or continuously with a scrubber device mounted within the photoreactor.
2. Scale-up is unreliable to a degree depending on the difficulty of modelling the very complex radiation field.

3.1.2   Annular Photoreactor

In this kind of reactor the reaction zone is delimited by two coaxial cylinders. The lamp is placed on the symmetry axis. It can operate in batch or continuous way. Practically all the photons emitted by the lamp reach the reacting medium. In the case of low thickness of suspension in which a relevant quantity of photons reach the outer wall of the photoreactor, an outer mirror can be used (Figure 6.5).

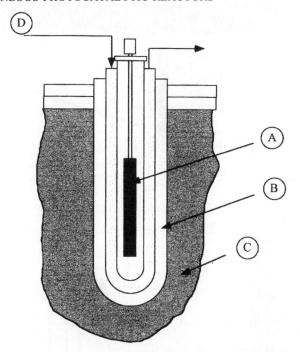

Figure 6.4. Immersion well photoreactor: A) light source; B) thermal insulation; C) reaction mixture; D) cooling liquid or filtering solution

3.1.3 Multilamp Photoreactor

This kind of photoreactor is cylindrical, externally surrounded by several lamps (Figure 6.6). The lamps are in turn surrounded by reflecting surfaces. This geometry is generally adopted when the lamps used are fluorescent tube ones, characterised by small power. Generally the reflecting surfaces have parabolic shape and the lamps are positioned on the focus.

3.1.4 Elliptical Photoreactor

In this configuration the reactor, of cylindrical shape, and the lamp are vertically positioned on the axes of the foci of a cylindrical reflecting surface of elliptical cross-section (Figure 6.7). Because of this geometry some of the photons impinge directly on the reactor wall but most of the photons emitted by the lamp reach the reacting medium radially. It must be noted that the radiant energy field at the outer wall of the photoreactor is not uniform and the intensity of the radiant field depends both on angular and axial coordinates.

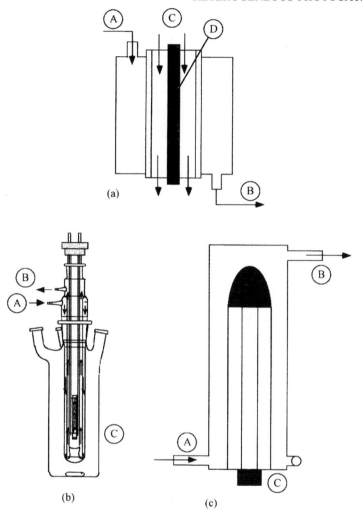

Figure 6.5. Annular photoreactors. (a) Continuous flow annular photoreactor: A) reactant inlet; B) product outlet; C) cooling liquid or filtering solution; D) lamp. (b) Batch annular photoreactor: A) cooling liquid or filtering solution inlet; B) cooling liquid or filtering solution outlet; C) lamp. (c) Fluidised bed annular photoreactor: A) tangential reactant inlet port; B) product outlet port; C) lamp

3.1.5 Film Type Photoreactor

In this kind of photoreactor the reacting medium forms a thin film of flowing liquid on the reactor wall which is externally irradiated. If the wall of the reactor is of cylindrical shape (case of falling film reactors) it may be designed for irradiation geometries with lamps external or internal to the reactor volume (Figure 6.8).

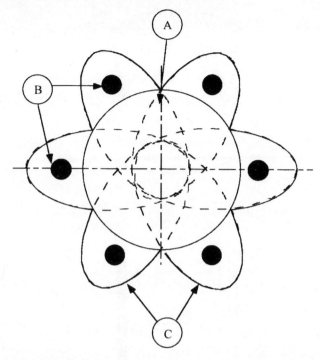

Figure 6.6. Multilamp photoreactor: A) reactor tube; B) extended light sources; C) parabolic reflectors

3.1.6 Flat Walls Photoreactor

When a parallel beam radiation field is available as in the case of short arc lamps provided with concave mirror and optics, or solar light irradiation the transparent wall of the reactor is usually flat (Figure 6.9(a, b)). The photoreactor shown in Figure 6.9(a) is called 'axially irradiated'. The radiation field can be 'cocurrent' when the radiation path has the same direction of reagent flow or 'countercurrent' in the other case. In the photoreactor shown in Figure 6.9(b) the radiation path is transversal with respect to the reagent flow. This type of arrangement is called 'cross-flow'.

3.2 SELECTION CRITERIA

The choice of the best photoreactor for a certain application must take into consideration some important points such as scale and purposes of the operation, available lamps, phases involved and thickness of the reacting medium. It is obvious that laboratory and industrial applications have different purposes. In the first case the target is to obtain reproducible results that have to be interpreted in

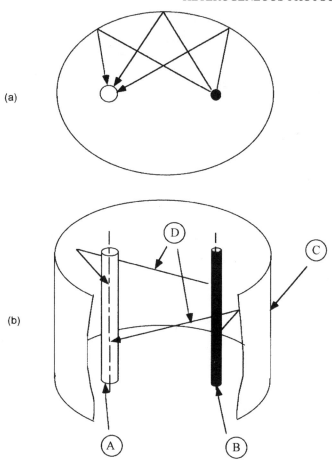

Figure 6.7. Elliptical photoreactor: (a) upward view; (b) perspective view: A) cylindrical reactor; B) lamp; C) elliptical reflecting chamber; D) emitted photons

terms of fundamental quantities, while in the last case the purpose is to obtain a certain product with minimum cost and maximum reliability. The lamps needed to carry out the process (number and geometry) often restrict photoreactor choice while when a solid phase is present, as in the case of photocatalysed reactions, to keep the solid phase in suspension mechanical agitation as well as fluidisation can be used. The optical thickness of the reacting medium is of great importance, in fact it gives an idea of the thickness of the layer of the irradiated medium. It is a guide for the choice of photoreactor type together with the knowledge of the average life of the intermediates.

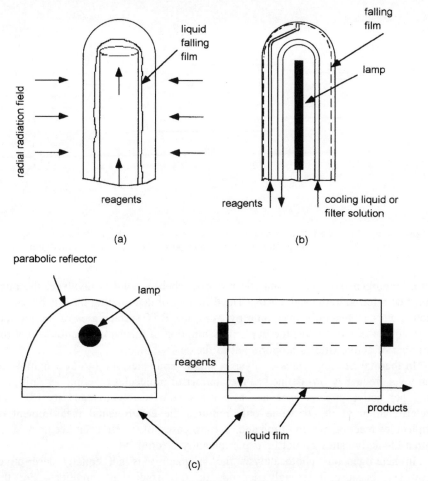

Figure 6.8. Film type photoreactors: (a) falling type photoreactor with positive geometry of irradiation (externally irradiated); (b) falling type photoreactor with negative geometry of irradiation (internally irradiated); (c) thin-film photoreactor

4   REACTIVITY PARAMETERS OF HETEROGENEOUS PHOTOCATALYTIC SYSTEMS

The most useful quantity to express the efficiency of a photocatalytic process is the quantum yield (Φ), as in photochemistry [7]. The quantum yield is defined as the ratio between the number of molecules transformed per unit time and the number of photons absorbed per unit time as follows:

$$\Phi = \frac{\text{transformed molecules/unit time}}{\text{absorbed photons/unit time}} \tag{6.1}$$

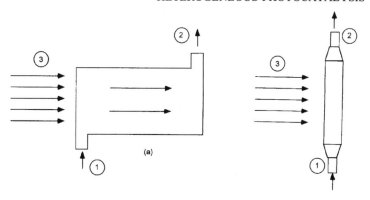

(b

Figure 6.9. Flat wall photoreactors: (a) axially irradiated photoreactor 'cocurrent arrangement'; (b) cross-flow photoreactor. (1) inlet flow; (2) outlet flow; (3) radiating beam

In a heterogeneous (photo)catalytic process, where a solid is involved, the rates must be referred to the active sites. The difficulty of determining the number of the active sites is generally overcome by using the BET surface area of the particle instead of the active sites; the implicit assumption of this simplification is that the number of active sites is proportional to the surface area.

In thermal heterogeneous catalysis a single parameter, the turnover number or turnover frequency (tn) defined as the number of molecules reacting per unit time and per active site, is used as basic measure of catalytic reactivity [8]. The determination of the tn value only requires the experimental measurement of molecules reacting per unit time and, in most cases, of the BET surface area, since often the active sites cannot be experimentally determined.

In heterogeneous photocatalysis the tn quantity is not equally descriptive, however, because it strongly depends on the irradiation conditions. For the occurrence of a photoreaction it is necessary that photons of suitable energy are absorbed by the semiconductor, which usually is in the form of polycrystalline porous particles. A first quantity that must be measured is therefore the rate of photon absorption (rpa), defined as

$$rpa = \frac{\text{absorbed photons}}{\text{time} \times \text{surface area}} \tag{6.2}$$

Once radiation is absorbed, electron–hole pairs are generated if the photon energy equals or exceeds the semiconductor band-gap energy. The pairs can: (i) generate thermal energy (recombination); (ii) determine a lattice conversion (photocorrosion); or (iii) be trapped by suitable surface species and initiate a reaction sequence with adsorbed surface molecules (photocatalysis). It must be noted that 'recombination' and 'photocorrosion' are mainly bulk phenomena while 'photocatalysis' is only a surface phenomenon.

On this ground a second quantity that must be measured is the specific reaction rate (srr), defined as

$$srr = \frac{\text{reacted molecules}}{\text{time} \times \text{surface area}} \tag{6.3}$$

Experimental determination of two quantities, the photon absorption rate and the specific reaction rate, is therefore needed for correctly evaluating the performance of heterogeneous photocatalytic systems.

The knowledge of rpa and srr parameters also gives information on the photon efficiency, i.e. on the quantum yield of the photoreaction.

In a homogeneous medium, Φ refers to the efficiency of a process in which the species that absorb photons themselves undergo chemical transformation. In a heterogeneous regime, on the other hand, the photons are absorbed by a species (the semiconductor), which plays the role of 'mediator catalyst'. As previously reported, upon photoexcitation the solid generates electrons and holes which eventually induce redox chemical reactions. In a heterogeneous regime, therefore, Φ has mainly the meaning of the evaluation of the efficiency of the process of energy transfer from a solid to a chemical product. Moreover it must be emphasised that, when the radiation impinges onto a surface of a solid, it is not totally absorbed, but it is part absorbed, part scattered and part transmitted. Indeed the quantum yield (Φ), defined as the ratio of reacted molecules to absorbed photons, gives the quantitative relationship between the number of molecules undergoing a particular process and the number of quanta absorbed by the system. The Φ values may be determined, knowing the rpa and srr parameters, by the following equation:

$$\Phi = \frac{srr}{rpa} \tag{6.4}$$

Since independent measurements of the absorbed photon flow and of the molecules reacted per unit time may be, in theory, performed for the same photoreacting system, it is easy to calculate the quantum yield of the photoreaction by directly applying the definition of Φ [9–11].

5 MODELLING

5.1 GENERALITIES

A model is an idealised representation of reality. It is a description of physical and chemical phenomena occurring in a process. The description can be based on either a microscopic or macroscopic view of the processes. Usually it is in the form of several mathematical equations or has the form of graphs or series of numerical data. As reality is complex in all circumstances, any model must always rely on a *priori* prediction of reactor behaviour. The series of equations usually written in the mathematical description of the processes occurring in a reactor correspond to several fundamental balances that apply conservation laws. In particular, if N different species are present, the set of equations comprises:

1. one differential mass balance for each species (N equations);
2. differential momentum balance;
3. differential energy balance;
4. 'constitutive'equations for transport phenomena, such as the Newton, Fourier and Fick laws;
5. all the equations relating the physico-chemical characteristics of the media (such as viscosity, density, diffusivity, absorbance, etc.) to temperature, pressure and concentration of the species present;
6. the kinetic equations for the various reactions taking place;
7. differential radiant energy balance for each wavelength and direction, that has to be added in the case of photon-involving processes, like photochemical ones.

This set of equations, together with the boundary conditions on each of the variables involved, describes any physico-chemical process taking place in a photoreactor.

A set of simplifying assumptions constitutes 'the model', and just as many simplifications should be made to solve the problem. The simplified model will not be able to reproduce reality in all aspects, but it will be successful if able to forecast with acceptable accuracy, just those parameters which are considered relevant for the case under examination.

5.2   RADIATION FIELD MODELS

In order to model the radiant energy field, some simplifying assumptions have to be made [12, 13]. In particular two different approaches have been assessed for the description of the lamp–reactor system:

1. The field of the radiation impinging the reactor is defined.
2. A model of light emission from the lamp is defined.

In the first case the energy radiant distribution outside the reactor transparent wall is known. It is the case, for example, of flat wall reactors, normally irradiated by parallel beams, in which the reagent flow is in the same direction. In this case making the plug flow assumption for the reagent stream, the model can be considered monodimensional. For reactors with cross-flow arrangement (see Figure 6.9(b)) at least two dimensions are needed to model the system. For cylindrical reactors the first simplification considers the radiation belonging to planes normal to the axis of the reactor (bidimensional radiation incidence model). If the beams are assumed to travel normally to the reactor wall as well, a simple radial incidence model is defined. The assumption that the beams impinge on the reactor wall in a diffuse way renders the model more realistic. In addition to this if the axial component of the radiation field is taken into account, a three-dimensional model is proposed, which is closer to the real situation, although much more complex. All the models rising from this approach correspond to the 'incidence models' reported in references [12, 13].

In the second approach the radiation field can be obtained by the defined model of light emission from the lamp. The most commonly used lamp geometries determine the lamp emission models which can be described as point source, line source and extended source. In the first, the light source is defined as a point emitting uniformly in all directions. Clearly this model applies quite well to very short arc lamps. In the second, which can be applied to long arc lamps, the radiation source is represented as a line. As in the case of the 'incidence' approach, the radiation can travel along planes or can be of the spherical or diffuse emission type. For both long arc lamps and fluorescent tubes the extended source model can be used, as the radiation can be considered as emitted from a cylindrical source. Moreover the photons can be emitted by the entire lamp volume or by the external lamp surface in either a spherical or diffuse way.

These three irradiation models correspond to the 'emission models' described in references [12, 13].

5.3 HETEROGENEOUS PHOTOREACTOR MODELLING

The main characteristic of heterogeneous photocatalytic processes is that a radiation field is present in the reaction system where, furthermore, absorption of photons by fine powders of semiconductor oxides takes place in order for the photoreaction to occur. On this ground photocatalytic processes can be considered as a case of radiative transfer in participating media [14, 15] and in their modelling the radiant energy balance equation must be added to the usual balance equations (mass, heat and momentum).

The radiant energy balance is generally performed in terms of photons, i.e. of particles of no mass and charge characterised by a wavelength, an associated energy, and a direction of propagation. By means of this balance the local rate of radiant energy absorption is known: this quantity is essential in the modelling of photocatalytic processes since it gives the concentration of the photons which are absorbed at any given position and time. It enters the constitutive equation for the mass generation rate of photochemically reacting species and therefore can be considered as the concentration of an additional reagent whose presence is necessary for the occurrence of the photoreaction.

If scattering and absorption phenomena are all of significance, the equation of transfer in a solid angle about a generic direction is:

intensity of radiation = loss by absorption + loss by scattering + gain by scattering

The situation of heterogeneous photocatalysis is that radiation absorption occurs by species which are not reacting but release the absorbed energy, thus causing the reaction to occur. As far as radiative transfer is concerned, the participating medium can be considered of constant composition and the radiant energy equation can be rather easily solved before considering the mass balance equations for the reacting species.

The availability of photons to be absorbed must be the highest one in order to exploit the whole reactor volume and to have an efficient photochemical process. Various phenomena may occur when incident radiation strikes a particle [16–18]; a part of the energy is removed by absorption and another part is redirected by scattering. For elastic scattering the photon energy and therefore the frequency of the photon is unchanged while for inelastic scattering the photon energy is changed. The scattering phenomena depend: (i) on the optical properties of the particle; (ii) on the size of the particle relative to the wavelength of the incident radiation; and (iii) on the particle geometry. For heterogeneous photocatalytic systems utilising fine powders at low concentrations, the scattering intensities from the individual particles can be added, thereby assuming that each particle scatters independently. It is also usually assumed that the medium does not enter into the optical behaviour of the medium–particle system.

The solution of the scattering behaviour provides very complicated relations for even simple particle geometries so that a number of simplifications are generally made [16, 17].

The first simplification is of letting the scattering particles be spheres. This is not a restrictive assumption; because the particles are in a random orientation, the angular distribution of scattered radiation viewed at a distance from the actual particles is the same as that scattered from spherical particles. A second simplification is to consider the limiting solutions for scattering from large and from small spheres. For large spheres the scattering is chiefly a reflection process and hence can be calculated from relatively simple geometrical reflection relations. For small spheres the approximation of Rayleigh scattering may be used. For the intermediate range, the general Mie scattering results can be used. Another simplification is to consider only diffuse surfaces. The particle surface can only act as a diffuse one, however, if particle dimensions are large compared with the wavelength of the incident radiation. It must be noted that only large particles (diameter greater than 50 μm) are of interest for realistic application in heterogeneous photocatalytic processes. For continuous stirred photoreactors working with liquid–solid suspensions the loss of catalyst by wash out must be avoided. Two ways can be arranged: the first one is to realise flow conditions so that solid particles remain inside the reactor and the second one is to have a separation apparatus (filtration, sedimentation, centrifugation, etc.) at the exit of the photoreactor. In both cases the presence of very small particles would render the whole operation very difficult and expensive.

In conclusion, in the case of heterogeneous systems the situation is much more complex due to the presence of more than one phase. Photoreactors working in gas–solid, gas–liquid and gas–liquid–solid regimes can be studied only experimentally. For the cases in which a model has been proposed, this model generally corresponds to the homogeneous system. Some approaches have been proposed to overcome this problem, but, as the model is much more complex in terms of mathematical description, the number of parameters that need to be measured increases, increasing in turn experimental difficulties.

6 CASE STUDIES

In this section we describe briefly the systems used for two photocatalytic processes, namely the photoreduction of N_2 to NH_3 in gas–solid regime, in the presence of H_2O vapour and the photo-oxidation of phenols in aqueous solution in a liquid–solid regime [19–21].

For the photo-reduction of N_2 the final product is NH_3 and the reactant H_2O is oxidized, while for the photo-oxidation of phenols (phenol, nitrophenols) the final product is CO_2 and the reactant O_2 is reduced. For the photo-reduction of N_2, two types of photoreactors were used. An elliptical photoreactor (Figure 6.7) and a continuous flow fixed bed photoreactor (Figure 6.10).

In the elliptical photoreactor a focus was occupied by the photoreactor and the other by a 400 W high pressure Hg lamp (Osram). The photoreactor was a fluidised

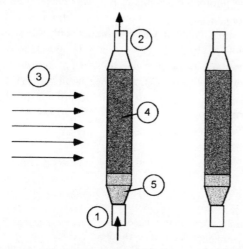

Figure 6.10. Continuous flow fixed bed photoreactor: (1) inlet flow; (2) outlet flow; (3) radiating beam; (4) catalyst bed; (5) glass pellets

one of cylindrical shape, made by Pyrex glass (i.d. $= 1$ cm). The other reactor was made by a Pyrex glass tube (i.d. $= 0.4$ cm; o.d. $= 0.6$ cm) filled with 1 g of the photocatalyst. It was irradiated by a 60 W high-pressure Hg lamp (Osram HWL).

Because the very low conversion, the use of the fluidised bed photoreactor (elliptical one) or of the fixed bed photoreactor does not influence the reactivity of the system. For this reason, the use of the gas-flow fixed bed photoreactor is preferred, because running this type of photoreactor is more easy than running the elliptical fluidised bed one, although the light distribution is better in the elliptical photoreactor than in the fixed bed one.

For the phenols photo-oxidation we used various type of photoreactors. In order to perform kinetics studies and intermediates determination, batch annular photoreactors with immersed lamp were used (Figure 6.5(b)). They are made by cylindrical vessels of Pyrex glass in which a medium-pressure Hg lamp is immersed. The reactor volumes used were 0.25, 0.5 and 1.5 l, while the lamp powers were 125, 500, 750 and 1000 W (Helios Italquartz).

For reactor modelling studies two types of continuous photoreactors were used. A fluidised annular photoreactor (Figure 6.5(c)) and a fixed bed flat photoreactor (Figure 6.3). The first was made by an annular Pyrex vessel irradiated from the inner part by a 500 W medium-pressure Hg lamp (Helios Italquartz). The lamp has a jacket in which a flow of distilled water was circulated in order to cool the lamp and to filter IR radiation. The solution was fed to the bottom and the resulting flow in the upward direction was sufficient to fluidise the catalyst. On the top of the photoreactor there was a cylindrical part that facilitated the settling of the particles to prevent powder loss with the exit flow.

The continuous flat photoreactor (Figure 6.3) was constituted by a thin layer of catalyst particles horizontally positioned on which a film of liquid solution flows in laminar regime. The open channel was in contact with the atmosphere and was irradiated by four 400 W medium-pressure Hg lamps (Helios Italquartz). The particular reactor geometry allows the use of sunlight and no separation or clarification operations are necessary. On the other hand the main drawback is the necessity of utilising several recycles of the liquid effluent to obtain a significant degradation of the pollutant.

REFERENCES

[1] M. Schiavello (Ed.), *Photocatalysis and Environment. Trends and Applications*, Kluwer, Dordrecht, 1988.

[2] M. Schiavello (Ed.), *Photoelectrochemistry, Photocatalysis and Photoreactors. Fundamentals and Developments*, Reidel, Dordrecht, 1985.

[3] E. Pelizzetti and N. Serpone (Eds.), *Photocatalysis. Fundamentals and Applications*, Wiley, New York, 1989.

[4] S. R. Morrison, *Electrochemistry at Semiconductor and Oxidized Metal Electrodes*, Plenum Press, New York, 1980.

[5] H. Gerischer, *Physical Chemistry: An Advanced Treatise*, Academic Press, New York, 1970.
[6] A. M. Braun, M. T. Maurette and E. Oliveros, *Photochemical Technology*, Wiley, New York, 1991.
[7] Glossary of terms used in photochemistry, *Pure & Applied Chem.*, **94**, 1990, 829.
[8] D. D. Eley, H. Pines and P. B. Weiss (Eds.), *Advances in Catalysis*, Academic Press, New York, 1977, Vol. 26.
[9] V. Augugliaro, L. Palmisano and M. Schiavello, *AIChE J.*, **37**, 1991, 1096.
[10] M. Schiavello, V. Augugliaro and L. Palmisano, *J. Catal.*, **127**, 1991, 332.
[11] V. Augugliaro, V. Loddo, L. Palmisano and M. Schiavello, *J. Catal.*, **153**, 1995, 32.
[12] O. M. Alfano, R. L. Romero and A. E. Cassano, *Chem. Engng Sci.*, **41**, 1986, 421.
[13] O. M. Alfano, R. L. Romero and A. E. Cassano, *Chem. Engng Sci.*, **41**, 1986, 1137.
[14] H. C. Hottel and A. F. Sarofim, *Radiative Transfer*, McGraw-Hill, New York, 1967.
[15] R. Siegal and J. R. Howell, *Thermal Radiation Heat Transfer*, McGraw-Hill, New York, 1972.
[16] H. C. van de Hulst, *Light Scattering by Small Particles*, Wiley, New York, 1957.
[17] A. E. Bohrens and D. R. Huffmann, *Absorption and Scattering of Light by Small Particles*, Wiley, New York, 1983.
[18] A. N. Matveev, *Optics*, MIR Publishers, Moscow, 1988.
[19] J. Soria, J. C. Conesa, V. Augugliaro, L. Palmisano, M. Schiavello and A. Sclafani, *J. Phys. Chem.* **95**, 1991, 274.
[20] V. Augugliaro, L. Palmisano, A. Sclafani, C. Minero and E. Pelizzetti, *Toxicol. Environ. Chem.* **16**, 1988, 89.
[21] V. Augugliaro, V. Loddo, G. Marcì, L. Palmisano and M. Schiavello, *Chem. Biochem. Eng. Quart.* **9**, 1995, 133.

Index

Index compiled by Geoffrey C. Jones

RETURN TO: CHEMISTRY LIBRARY
100 Hildebrand Hall • 510-642-3753

LOAN PERIOD	1	2 1 Month	3
4		5	6

ALL BOOKS MAY BE RECALLED AFTER 7 DAYS.
Renewals may be requested by phone or, using GLADIS, type inv
followed by your patron ID number.

DUE AS STAMPED BELOW.

SEP 08		

FORM NO. DD 10   UNIVERSITY OF CALIFORNIA, BERKELEY
3M 7-08   Berkeley, California 94720–6000